ICME-13 Topical Surveys

Series editor

Gabriele Kaiser, Faculty of Education, University of Hamburg, Hamburg, Germany

More information about this series at http://www.springer.com/series/14352

Florence Mihaela Singer
Linda Jensen Sheffield · Viktor Freiman
Matthias Brandl

Research On and Activities For Mathematically Gifted Students

 Springer Open

Florence Mihaela Singer
University of Ploiesti
Ploiesti
Romania

Linda Jensen Sheffield
Northern Kentucky University
Highland Heights, KY
USA

Viktor Freiman
Faculté des sciences de l'éducation
Université de Moncton
Moncton, NB
Canada

Matthias Brandl
Didaktik der Mathematik
Universität Passau
Passau, Bayern
Germany

ISSN 2366-5947   ISSN 2366-5955 (electronic)
ICME-13 Topical Surveys
ISBN 978-3-319-39449-7   ISBN 978-3-319-39450-3   (eBook)
DOI 10.1007/978-3-319-39450-3

Library of Congress Control Number: 2016940119

Printed on acid-free paper

This Springer imprint is published by Springer Nature
The registered company is Springer International Publishing AG Switzerland

Main Topics You Can Find in This ICME-13 Topical Survey

- Nature of Mathematical Giftedness
- Mathematical Promise in Students of Various Ages
- Research into Practice: Pedagogy, Programs and Teacher Education

Contents

Research On and Activities
For Mathematically Gifted Students

1 Introduction

In 1980, *An Agenda for Action: Recommendations for School Mathematics for the 1980s* from the National Council of Teachers of Mathematics in the United States noted that "The student most neglected, in terms of realizing full potential, is the gifted student of mathematics. Outstanding mathematical ability is a precious societal resource, sorely needed to maintain leadership in a technological world" (NCTM 1980, p. 18). Over 35 years later, the world has certainly become more technological. In this Topical Survey, we explore whether our gifted mathematics students around the world are closer to realizing their full potential and suggest strategies and needed research to make that happen.

There is a continuous debate around the conception of giftedness and its definition. Over the time and places, several terms are being used in the context of gifted learners: mathematically gifted and talented, (highly) able, (intellectually) precocious, bright, mathematically advanced, among many others. While we use many of these terms in our survey, reflecting choices made by the researchers and practitioners who have contributed to the field, the term 'mathematically promising' introduces by the NCTM Task Force in the mid-90s, seems to us the most appropriate to grasp the complexity of the topic in its largest and broader sense.

Because the domain of mathematical giftedness is, as an interdisciplinary domain, still under development, we organize our discourse based on a list of questions to which we try to give evidence-based answers. Examples drawn from longitudinal studies, micro-analyses of classroom interactions, various educational programs and projects, along with findings from recent cognitive and neuroscience studies that offer insights into how the mathematically promising mind works are brought together to offer a synthesis of state of the art on research in the education of the gifted. We hope that this approach will stimulate a constructive debate and will lead to new international research and development in this emerging field.

© The Author(s) 2016   1
F.M. Singer et al., *Research On and Activities For Mathematically Gifted Students*,
ICME-13 Topical Surveys, DOI 10.1007/978-3-319-39450-3_1

2 Nature of Mathematical Giftedness

There is a common truism to say that there is no single definition of mathematical giftedness. For this reason, we do not give much place to present theoretical definitions, we just restrain to a minimum needed for understanding the topic; we mostly focus on results from empirical research. Different actors involved in the educational systems directly or indirectly, as teachers, parents, researchers, or students have different views on giftedness. We explore these different views through various lenses, and the first question we address in this respect is:

2.1 What Is Mathematical Giftedness?

To answer, we start with a few words about general giftedness, and then we focus on what is specific to mathematics.

2.1.1 General Giftedness

There are numerous definitions of a gifted child. Some emphasize the child's current level of achievement based on an overlap and interaction among three clusters of traits: above average ability, task commitment, and creativity (Renzulli 1986); whereas for others, the key is the child's potential to perform at a level significantly beyond age-peers (e.g. Gagné 2003). According to Gagné (2003), personal characteristics such as motivation and temperament, as well as environment and the interplay between these aspects and innate giftedness, play an important role in the development of talent. Thus, certain traits should be evident in potentially gifted young children, while others are developed through instruction. Therefore, we are talking about a hidden potential with a genetic component of disposition on the one side, versus observable performance or achievement or expertise on the other side. Thus, the talent development process is a progressive transformation of gifts into talents.

There exists now extensive literature on the identification of gifted children. In terms of cognitive behaviors, a fast pace of learning, exceptional memory, extended concentration span, ability to understand complex concepts, enhanced observational ability, curiosity, and an advanced sense of humor should be apparent (e.g. Harrison 2003). Most research in giftedness has so far concentrated on intellectual and academic aspects. However, high intellect and creativity are frequently accompanied by personality factors that impact the life of gifted children, such as: advanced moral judgment; heightened self-awareness; heightened sensitivity to the expectations and feelings of others; perfectionism; introversion; high expectations of self and others; idealism and a sense of justice; and higher levels of emotional depth and intensity (e.g. Winner 2000; Clark 2002).

2.1.2 Mathematical Giftedness

Mathematical giftedness is sometimes seen as a specific part or kind of giftedness. However, narrowing the focus from the look on general giftedness to the subset of mathematical giftedness does not necessarily imply taking only into account a subset of the items describing general giftedness. When discussing mathematical giftedness, many complexities occur and some domain-specific aspects are to be emphasized.

According to Krutetskii, "mathematical giftedness" is the name given to *a unique aggregate of mathematical abilities that opens up the possibility of successful performance in mathematical activity* (Krutetskii 1976, p. 77). His comprehensive investigation of mathematical ability was designed to explore the nature and structure of mathematical abilities over a 12-year period. He defined ability as a personal trait that enables one to perform a given task rapidly and well, and contrasts this to a habit or skill, which relates to the qualities or features of the activity a person is carrying out. Krutetskii (1976) uses the term 'mathematical cast of mind' to describe the mathematically gifted students' tendency to view the world through a mathematical lens. That means that gifted and talented mathematics students have, among other capacities, the ability for rapid and broad generalization of mathematical relations and operations, and flexibility of mental processes.

Given the domain-specificity of mathematical giftedness, it always implies a collection of certain mathematical abilities and personal qualities. However, looking at the subject at different times and within different cultural contexts we get changing definitions corresponding to the "law of cultural differentiation" (Irvine and Berry 1988). Furthermore, talking about mathematical giftedness is unavoidably tied to talking about mathematics. In complex-system theoretical terms, the open and viable mental construct "mathematical giftedness" is connected to its environmental system "mathematics" consisting both of the mathematical truths (theorems, definitions, axioms, etc.) and the individual researchers making these up. Now, different notions of mathematics determine different concepts of mathematical giftedness, or as Freudenthal put it, "The definition of mathematics varies. Each generation and each subtle mathematician within each generation formulates a definition that corresponds to his or her skills and insights." (quoted in Käpnick 1998, p. 53). Hence, predominant philosophical notions about mathematics influence a conception of mathematical giftedness, which means there is a structural connection between mathematical giftedness and mathematics itself.

Lists representing necessary aspects of mathematical giftedness, generally split into abilities specific to mathematics on the one hand (such as: mathematical sensibility, exceptional memory, rapid content mastery and structuring, atypical problem solution, preference for abstraction, interest and enjoyment of mathematics, success in identifying patterns and relationships, lengthy concentration span, generalizing and reversion of mathematical processes) and general personality traits on the other (intellectual curiosity, willingness of exertion, joy and interest in problem solving, perseverance and frustration tolerance, ability to engage in independent self-directed activities, and affinity for challenging tasks).

Giftedness then can manifest as school giftedness and creative productive gift-edness. The first is manifested in the facility to take standardized tests and acquire knowledge, while the second refers to the ability to create new products or processes. In addition, the term good student might be used to describe a high achiever who is not gifted and who is often focused on pleasing teachers or parents (e.g. Brandl 2011). Being gifted in mathematics does not necessarily lead to high attainment in this subject while high attainment in mathematics does not necessarily mean being mathematically gifted (e.g. Szabo 2015; Brandl and Barthel 2012; Brandl 2011; Öystein 2011). Hong and Akui (2004) offer a similar distinction when they introduce the constructs of academically gifted and creatively talented students. The ones in the first category are those performing well in "school" mathematics, the high achievers, while the other ones are highly interested and active individuals, yet they are not high achievers. It seems that the category doing both is less visible in school settings.

2.2 A Discovery or a Creation?

Differing definitions of mathematics as a domain of knowledge, on the one hand, and mathematically gifted, on the other, may lead to different responses to this question. We will approach both aspects, showing some strengths and weaknesses of each perspective. Research focusing the question if mathematical giftedness is a discovery or a creation can have a major impact in improving the teaching strategies that address the development of mathematical abilities in students of various ages.

2.2.1 Mathematical Giftedness as a Discovery

There is a largely accepted assumption that giftedness might equal intelligence that is measurable by IQ-tests. From here comes the belief that about 2 % of people can be seen as (highly) gifted. However, there are various problems related to considering IQ tests as unique reference for identifying giftedness. To list just a few critiques: the IQ-score is not stable or constant over time but changes in connection to the further development of the tested person. In addition, there is not a single IQ-test, but different IQ-tests that collect different data and make different statements, as they are designed for specific norm groups. More than one hundred years ago, Binet (1909), who is often acknowledged as the inventor of the modern intelligence test, protested against what he termed the brutal pessimism of philosophers who asserted that an individual's intelligence is a fixed quantity, a quantity that cannot be increased. Binet felt that with practice and training one might become more intelligent than he/she was before. He claimed that not always the people who start out the smartest would end up the smartest.

Features of mathematical ability at an early age. Research literature in this area describes various early signs of mathematical giftedness in children (e.g. Diezmann and Watters 2000). In a 2-year study that examined 15 mathematically gifted and talented students aged from 10 to 13 years, Bicknell (2008) characterizes mathematical giftedness through the eyes of parents, students, and teachers. Most of the parents recognized their child's abilities in mathematics at an early age. The parents' descriptions of their children at pre-school age give an image of what might be seen as innate abilities in these children. The characteristics identified by parents include "impressive concentration" and the ability to work independently for a relatively long period of time on a particular task. Young children no older than 2–3 years were self-initiating games involving numbers and numerical patterns, showing a real fascination for numbers and how they behave in mathematics operations. The types of activities the parents observed in their children at an early age indicated an interest in mathematically driven games, such as: building with construction blocks, creating symmetrical patterns, ordering objects, and completing puzzles and jigsaws in unconventional ways, spending hours of concentration in such activities. Others made connections between ballet movements and angles in geometrical rotations, or have shown a relatively sound concept of number and in some cases an interest in concepts such as time and space (Bicknell 2008).

But not all young potentially gifted children show these signs of mathematical giftedness. In a questionnaire given to more than 100 parents of mathematically gifted students most of the parents' answers support these findings; however, some pointed out that they did not recognize signs of giftedness before their child began school (Nolte 2012). Moreover, descriptions of mathematical precocity in 2- or 3-year old children should not necessarily lead to the idea that their mathematical abilities were innate. These could have been due to parental and other environmental factors.

Features of mathematical ability at the school level. Once gifted children began school, their level of interest and ability in mathematics compared to their peers became more apparent. The teachers observed in these children the different pace of mathematics learning, an intuitive mathematical knowledge in problem solving, their keen interest in mathematics, the sense of humor and ability to think in more abstract terms than their age peers, as well as more mental flexibility and a discourse based on logical thinking. Perseverance and excitement with mathematical problems were also observed. Other aspects of mathematics that, according to the students, confirmed their mathematical giftedness include success in competitions; competence with basic mathematical facts; speed of computational skills; problem solving abilities; capacity to work on 'special projects', or on more/different work (than their classmates) to complete independently (e.g. Bicknell 2008; Subotnik et al. 2012). The conclusions should be nuanced however, because not all of the students identified as mathematically gifted would categorize themselves as gifted, and they even would not rate mathematics as their favorite subject.

Sometimes students' abilities manifest quite differently across mathematical domains and both students and teachers have recognized these differences. For example, some students are stronger in visual patterns, and transformational geometry involving rotation or translation, while some others have good mathematical computational skills. However, this might not be directly related to mathematical giftedness, but to different cognitive styles. In a study investigating the relationship between three ability-based cognitive styles (verbal deductive, spatial imagery, and object imagery) and performance on geometry problems that provided different types of clues, Anderson et al. (2008) found both spatial imagery and verbal reasoning cognitive styles were helpful in solving some types of geometry problems, (but not object imagery, which has been found to relate more with art creativity).

The existence of a specific individual giftedness potential is not sufficient for high performance as the phenomenon of "underachiever" shows. Formal or informal learning provides a means of transforming this potential into talents or systematically trained abilities (achievement). Nevertheless, some researchers claim this potential is necessary for excellent results in assessment (e.g. Heller and Ziegler 2007; Bicknell 2008).

Features of mathematical ability at the university level. Research on university gifted students is quite limited. A possible explanation is that gifted students might have learned to hide their giftedness and, thus, it might be difficult to identify them at the university level (Albon and Jewels 2008). A different explanation might be that IQs do not remain stable over time and the fact that the interplay between interests, activities, environment, and mathematical explorations affect students' mathematical achievement leads to question whether there is a need to distinguish between giftedness and expertise as students enter the university level and beyond.

One area that has been studied is that of students entering college or university at a younger age compared to their colleagues. A survey of empirical research shows that, in general, early entrants earn higher grade point averages than regular students, are more likely to graduate, and are likely to earn other academic honors and pursue graduate studies (Olszewski-Kubilius 2013).

Adult mathematical achievement. Those gifted children most likely to develop their talent to the level of an expert will be those who have high drive and the ability to focus and derive flow from their work, those who grow up in families that combine stimulation with support; and those who are fortunate to have inspiring teachers, mentors and role models. Those gifted children often discover their talent in adulthood when they are catalyzed by a crystallizing experience, a life-changing event in which a gift is discovered and self-doubts are dispelled (e.g. Winner 2000).

2.2.2 Mathematical Giftedness as a Creation

What does influence the development of abilities and performance? Are some children born with special characteristics that allow them to become mathematically gifted, or is mathematical talent and expertise something that can be developed or created in the large majority of both male and female students from all ethnic and socio-economic groups regardless of traits inherited at birth? A term that gains more terrain is *mathematical promise*, which has been developed by the National Council of Teachers of Mathematics (NCTM) as a function of maximizing variables such as abilities, motivation, beliefs, and experiences or opportunities (Sheffield et al. 1999). NCTM (1995) suggested using the term promising rather than gifted, purposely broadening the definition to include a much greater range of students and to open the possibility of creating students with outstanding mathematical abilities and not simply identifying students with mathematical pre-existing expertise and passion (Sheffield et al. 1999). Gagné (e.g. 2003) talks about having "gifts" as a prerequisite for developing talents while others, including those that favor the definition of mathematical promise do not.

Dweck (2006) has shown that middle grades students who believe in a "fixed" mindset, that is a belief that they are born with certain "fixed" abilities, do more poorly in learning mathematics than those who believe in a "growth" mindset, understanding that their brain changes and develops. Those last students were more challenged to learn and more successful in middle grade mathematics. This is true for gifted students who believe that they have a math brain as well as those who believe that they do not. In discussing the "myth of the mathematically gifted child", Boaler (2015) makes a powerful case for the harm caused by this idea of genetic determinism and teaching mathematics as a subject that is used to separate children into those who have the math gene and those who don't. Instead, she calls for teaching mathematics as a lens to view the world that is available to all students through study and hard work.

Furthermore, it holds that students who are interested in mathematics will be more likely to develop mathematical talents. Kruteskii (1976) also stressed the necessity of an interest in mathematics in order to be successful in this subject, and "if the teacher is able to awaken his interest in it and his inclination to study it, that pupil 'carried away' by mathematics can quickly achieve great success" (Kruteskii 1976, p. 347). Some preliminary results from a look into different fostering settings of mathematically gifted students also indicate that the "perspective of mathematics of somebody who is interested in mathematics differs essentially from the one averaged over an "ordinary" class; it seems also to be more positive than that one averaged over a "high-attaining" class." (Brandl 2014, p. 1162). On the contrary, by choosing subgroups according to typical characteristics of mathematical giftedness from Tall's (2008) formal-axiomatic-world (e.g. deep interest in mathematics, inclination for the beauty of mathematics, loving to play around within elements of mathematics voluntarily) out of a sample of very high attaining students, significant correlations between the different subgroups' fields of mathematical interests have been found (Brandl and Barthel 2012). Additionally, the members of these subgroups who manifested special interests represented the top part of the sample, considering their marks in mathematics.

2.3   What Theoretical Frameworks and Methodologies Are Helpful?

Development models for mathematical giftedness are, in some cases, inspired by development models for general giftedness such as Gagné's Differentiated Model of

Giftedness and Talent (Gagné 2009), Renzulli's tripartite model, in which gifted-ness is the product of three interacting clusters of traits: above average intellectual ability, high levels of creativity and high levels of task commitment (Renzulli 1986), Ziegler's Actiotope Model of Giftedness (Ziegler 2005) or Heller's Munich Model of Giftedness (e.g. Heller and Ziegler 2007). Dai (2010) scrutinized some of the held assumptions about the nature of giftedness and explained why a contextual, developmental framework of approaching giftedness is a more viable alternative to the traditional psychometric framework.

As other researchers in this area (e.g. Leikin 2011), Szabo (2015) claims that, in the last decade, only a few studies are focused on analyzing traits of mathematically gifted and high achieving students in a conceptual perspective. Similarly, very few studies analyze the connection between those students' biological and cognitive capacities and their mathematical performance. More research is needed on theo-retical frameworks or models for explaining mathematical giftedness and promise. In the meanwhile, from a pragmatic view, in some countries, various institutions, NGOs or communities of parents or teachers included, started to develop general frameworks for designing differentiated learning experiences for gifted students, which have the role to complement the official standards and benchmarks.

3 Mathematical Promise in Students of Various Ages

As noted earlier, not all high achievers are gifted and not all gifted children are high achievers. Just as students with identified gifts may not be high performers, high attaining students do not need to be (highly) gifted (e.g. Brandl and Barthel 2012; Öystein 2011). For example, depending on the school environment, students might receive very good grades in examinations, but examination tasks may be only aimed at computational or algorithmic abilities and not demanding any type of non-routine or innovative approach. This may not mean that the successful student is a gifted or a talented future mathematician.

In her recent book, Boaler (2015) explains that teaching needs to reflect the new science of the brain and communicate that everyone has the potential to learn mathematics effectively, not only those believed to hold a "gift". From the same perspective, some authors try to identify profiles or traits of mathematically promising students (Budak 2012; Printer et al. 2015), focus the identification of such students (Vilkomir and O'Donoghue 2009), or discuss how to foster the promise of high achieving mathematics students (Hoeflinger 1998; Zmood 2014). We use the concept of *mathematically promising students* in order to cover a large span of baseline abilities, backgrounds, and environments as well as to capture the vision that most students have capabilities that, with adequate training, can help them reach much higher levels of mathematics performance.

3.1 How Are Cognitive, Social, and Affective Aspects Connected?

Traditionally, the identification of gifted students has been linked to intelligence tests and consequently, the identification of mathematically gifted students has often built on the general giftedness identification. Many teachers use checklists of typical traits to identify these students. Still, others focus on the development of mathematical talent rather than its identification.

3.1.1 Cognitive Indicators

As noted earlier, there are difficulties in assuming that giftedness equals intelligence as measured by IQ tests. Subsequently, researchers have tried to describe the complex features that could define domain-specific giftedness.

Following up on Kruteskii's work, many authors and research teams developed lists of cognitive characteristics of a gifted child (e.g. Diezmann and Watters 2000). An example of an age-group-specific list is the one from Käpnick (1998) who investigated primary school children's characteristics of giftedness: remembering mathematical facts; structuring mathematical facts; mathematical sensitivity and mathematical fantasy; transferring mathematical structures; intermodal transfer; and reversing lines of thought. Similarly, but more detailed aspects taken from a three-dimensional mathematical giftedness pattern represent the guideline for a process-based identification of mathematically gifted third and fourth graders during regular lessons, in Winkler and Brandl (2016). Assmus (2016) tested Käpnick's items with second graders and found the following characteristics of mathematical giftedness in early primary school children: ability to memorize mathematical issues by drawing on identified structures, ability to construct and use mathematical structures, ability to switch between modes of representation, ability to reverse lines of thought, ability to capture complex structures and work with them, ability to construct and use mathematical analogies, mathematical sensitivity, and mathematical creativity.

Other indicators of mathematical giftedness may include: unusual curiosity about numbers and mathematical information; ability to understand and rapidly apply mathematical concepts; high ability to identify patterns and to think abstractly; flexibility and creativity in approaching problem solving; ability to transfer mathematical concepts to an unfamiliar situation; persistency and resilience in solving challenging problems (Stepanek 1999). Sriraman (2005) focused on mathematical processes through which various authors define mathematical giftedness at K-12 level. These processes include, among others, the ability to: abstract, generalize, and discern mathematical structures; manage data; master principles of logical thinking and inference; think analogically and heuristically; visualize problems and/or relations; distinguish between empirical and theoretical principles; think recursively. However, researchers note that these indicators should not be used as rules for qualifying students as being mathematically gifted. Not every mathematically gifted student will display all these characteristics, or they may emerge at different times depending on the student's development. Much of identifying gifted students relies on ongoing assessments and teacher observations, as well as the level of problems against which students show these traits (Nolte 2012).

3.1.2 Social and Affective Indicators

A combination of internal and situational factors may cause problems for mathematically promising students. Among the issues that may affect them are *their asynchronous development, their socialization problems, and their own problems with self-learning.*

Asynchronous Development in Gifted Children. A common issue with gifted children is asynchronous development, i.e. non-uniform development through the intellectual, emotional, social, and physical domains. For example, a student whose mathematical abilities are far beyond other areas of development may have difficulties in adaptation to situational contexts. Suggestions for programs for these students will be explored in the next section.

Need for Social Acceptance. In the absence of understanding and support, highly gifted children may find themselves very different from their age mates and may face negative reactions in situations where conformity is valued. Many learn to mask their abilities in order to relieve their social problems (Gross 2003), especially in areas where mathematical expertise is not valued. This could hinder the further development of their unique abilities, and lead to a loss of self-esteem. However, a large body of literature has found gifted children to be superior not only intellectually, but also physically, emotionally, and socially (e.g. Cross et al. 2008).

Problems with Self-Learning. Gifted children are often inclined to learn things on their own, and are tempted to solve by novel methods problems that may be beyond their current abilities, introducing large amounts of error and frustration. Unassisted, such children may down-regulate their ambitions, develop a fear of making mistakes, and reduce productive risk-taking behaviors (Freehill 1961).

Twice-Exceptional Children are those who possess giftedness or exceptional ability in one or more areas in combination with special needs, a learning disability or other handicap in other areas. They may achieve high scores on certain intelligence tests but may not do well in school. These children represent a category among the gifted that is especially at risk without knowledgeable intervention. Self-esteem issues are often disproportionately high in children with learning disabilities or with notable asynchronous development, as they tend to judge themselves by what they cannot do rather than by what they can. This problem is relieved somewhat by sharing with them assessments of their abilities so that they develop more appropriate levels of self-esteem (e.g. Nolte 2013; Nordheimer and Brandl 2016).

3.2 What Does Brain Research Say?

Mathematical giftedness started to be conceptualized in the recent decades within a context that is sensitive to modern biology. The findings from educational neuroscience help understanding how gifted students might better be taught and helped to plenarily develop.

The application of neuroscience research to mathematically gifted students is somewhat controversial, however, and a deeper analysis is needed to allow drawing more detailed and specific conclusions. Reviewing studies of patient calculation based on magnetic resonance imaging, Dehaene et al. (2003) have proposed that specific regions of the brain play distinct functional roles in arithmetic. Several mathematical concepts, including number sense, are constructed based on spatial-numerical mapping (briefly, associating numbers with positions on a line). It was assumed that this association is a basic cornerstone for arithmetic skills, but recently, Cipora et al. (2015) concluded that the relationship between spatial-numerical associations and arithmetic skills are rather weak or caused by mediating variables. Nevertheless, interventions based on relationships between space and numbers can be beneficial for arithmetic skills because space is a powerful tool to understand arithmetic concepts.

Synthesizing some recent studies, Geake (2006) inferred that doing mathematics critically involves the lateral frontal cortices to support working memory; the temporal cortices (and hippocampus) to reconstruct knowledge from long term memory; the orbitofrontal cortices and the anterior cingulate for decision making, in turn mediated by regions within the limbic sub-cortex; areas of the fusiform gyri and temporal lobes for sequencing of symbolic representations; the parietal lobes for spatial reasoning about conceptual inter-relationships; and the cerebellum for mental rehearsal.

There have been several neuroimaging studies of the brain function of mathematically gifted children compared with normal age-matched peers. O'Boyle et al. (2005) in a fMRI study found areas of the brain that were involved in both pre-algebraic and geometrical thinking of able young mathematicians. It seems that mathematical thinking requires the coordinated participation of several neural systems, which in the brains of gifted mathematicians seem more extensive throughout both right and left hemispheres (e.g. Geake 2009; O'Boyle et al. 2005).

Data derived from several psychophysiological studies support an important relationship between the specialized capacities of the right hemisphere and mathematical ability, but this may depend on the type and complexity of the task (e.g. Jin et al. 2007). However, fMRI studies comparing the functioning of gifted versus other brains (Lee et al. 2006) showed that gifted individuals did not use more, or different, brain structures; rather, increased activation of the entire frontal-parietal network was noted, perhaps indicating higher-than-average activity distributed across the brain when performing difficult tasks. There are different findings depending on the difficulty of the problem. If the problem is easy, there is less activation of the frontal lobe. The discussion is sometimes in terms of 'neural efficiency', where gifted functioning involves a more integrated brain with greater cooperation between the hemispheres (O'Boyle 2008), with reduced activity in certain areas as compared with average brains when performing similar tasks—possibly implying that gifted brains spend less time on such tasks.

Mathematical giftedness as a form of intelligence related to enhanced mathematical reasoning has been tested using a variety of numerical and spatial tasks. For example, gifted adolescents displayed enhanced connectivity patterns during a task involving mental rotation of complex three-dimensional block figures when compared to an average control group. Findings are consistent with previous studies linking increased activation of the frontal and parietal regions with high fluid intelligence, and may be a unique neural characteristic of the mathematically gifted brain, at least for this type of task. As mentioned in the previous section, Anderson et al. (2008) found that both spatial imagery and verbal deductive cognitive abilities were important for solving geometry problems, whereas object imagery was not. This correlates with the observation that object imagery is important in arts professions, while spatial imagery is helpful in math, science and engineering.

A number of studies investigating the brain characteristics of mathematically gifted youth indicate that they possess different functional organization as compared to those of average mathematics ability (O'Boyle et al. 2005; O'Boyle 2005; Raghubar et al. 2010). Specifically, data from a variety of behavioral and psychophysiological experiments tend to suggest enhanced processing reliance on the right cerebral hemisphere and heightened interhemispheric communication, as unique functional characteristics of the mathematically gifted brain. Notably, these brain differences may have important implications for the nature and timing of mathematics instruction.

Case studies of extremely gifted individuals often reveal unique patterns of intellectual precocity and related brain activity. Presenti et al. (2001) using PET measures and brain imaging, found calculation in an adult mathematical prodigy (Rüdiger Gamm), to be uniquely mediated by right prefrontal and right medial temporal cortex. A long chain of arithmetical operations and data handling would put a considerable strain on normal working memory, yet many types of experts show increased capacities for the temporary storage of task-relevant materials.

In another study, comparing memory and speed of processing in 160 16- to 18-year-old general gifted and excelling in mathematics male students, Leikin et al. (2013) examined the memory and speed of processing abilities associated with general giftedness (G) and excellence in mathematics (E). Working-memory was found to be related to both G and E factors. The results reveal that G factor is related to high short term memory and that E factor is associated with high visual-spatial memory. Gifted students who excel in mathematics (G-E group) outperformed all in speed of processing tasks. The findings of this study partly support previous observations and suggest that memory and speed of processing abilities seem to be important factors in explaining mathematical giftedness. With this in mind, educational programs for G-E students, should address the observation that these students have high abilities in visual-spatial memory and in information processing and implement the use of visual aids in teaching mathematics in gifted classes (e.g. Leikin et al. 2013).

3.3 What Are the Differences Between Mathematical Novices and Experts?

Mathematical expertise implies the existence and use of two types of knowledge: explicit knowledge of facts, principia, formulae pertaining to the domain, and implicit knowledge of how to operate with them, i.e. declarative and procedural knowledge. Research on the cognitive sub-processes involved in the expert problem solving of the gifted, as compared to the problem solving of the average person, has attributed the difference between these two populations to selectivity in their encoding, comparison and combination sub-processes. Gorodetsky and Klavirb (2003) extend this list by adding two sub-processes that are imported from the literature on experts and novices: namely, retrieval and goal directness. Based on these five sub-processes, middle high school students (gifted and average) solved insight problems, without and with analogical learning, and were asked to report on the solution process they undertook. Though both the gifted and the average were able to arrive at correct solutions, the study shows that they employed different sub-processes in doing so (Gorodetsky and Klavirb 2003).

Usiskin (2000) has devised an eight-tiered hierarchy, which ranges from Level 0 to Level 7, to classify mathematical talent. In this hierarchy, Level 0 (No Talent) represents adults who know very little mathematics, and Level 1 (Culture level) represents adults who have rudimentary number sense as a function of cultural usage with mathematical knowledge comparable to that of students in grades 6–9. Clearly, a very large proportion of the general population would fall into the first two levels. Thus, the remaining population is spread throughout Levels 2 through 7 on the basis of mathematical talent, from Level 2 representing the honor high school student who is capable of majoring in mathematics, up to Level 7 with the Fields Medal winners in mathematics, or geniuses like Leonard Euler, Karl Friedrich Gauss, Srinivasa Ramanujan and others (Usiskin 2000).

While Usiskin's levels start from a social framing, Glaser (1988) characterizes expertise on six cognitive dimensions: knowledge organization, complexity of problem-solving representation, goal-oriented procedural knowledge, automatic procedures, and metacognition. The model of the gifted and talented learner as an expert knower and thinker can be used to differentiate the regular curriculum in the sense that the transition from the novice to expert knower can be mediated by adequate strategies and resources the teachers of the gifted are supposed to organize and develop. Still, more research is needed for identifying evidence-based pathways that lead to increasing expertise of mathematically promising students.

3.4 How Are Mathematical Creativity and Giftedness Connected?

The discussion about the differences between novices and experts cannot avoid the relationship between *expertise and creativity*. There are conflicting views about this relationship. Analyzing the students' level of expertise may depend on how expertise and creativity are defined. Thus, for example, Diezmann and Watters (2000) claim that for a student to be creative, he/she needs some intellectual autonomy and expertise. Expertise is therefore seen as a necessary precondition for the manifestation of creativity. On the other hand, Craft (2005) claims that every student is capable of creative manifestations regardless of level of expertise.

In a recent study focused on how expertise interacts with creativity in problem solving and posing, Singer and Voica analyzed the results of activities undertaken by mathematics students enrolled in a pre-service teacher-training program and found that, in the process of problem solving and problem posing, expertise and creativity support and mutually develop each other. Consequently, a possible method of training excelling students is through practicing tasks appropriate to their level of mathematical abilities, but containing nonstandard challenging components for which that person does not have yet internalized models of solving, in order to train metacognitive self-regulation capabilities through creative leaps. The authors revealed that, in the process of building a solution for a nonstandard problem, expertise and creativity interact and enable bridges to the unknown, mutually developing each other. This interaction leads to an increase in students' expertise (Singer and Voica 2016).

In professional mathematics, "creative" mathematicians constitute a very small subset within the field. From the hierarchical classification of mathematical talent outlined by Usiskin, it appears that in the professional realm, mathematical creativity implies mathematical giftedness, but the reverse is not necessarily true. Usiskin emphasized that students have the potential of moving up into the professional realm (Level 5) with appropriate affective and instructional scaffolding as they progress beyond K–12 into the university setting.

Hoyles (2001) analyzed the role that a computer-based learning environment can play in the navigation between skills and creativity in teaching mathematics. She concluded that technology-based inquiry opens opportunities for the advancement of students' mathematical creativity.

Much of the empirical research explores the learning processes of mathematically talented students through problem-solving strategies, but problem posing has also been linked to mathematical creativity. As early as 1973, Jensen (Sheffield) studied relationships among numerical aptitude, mathematical creativity, and mathematical achievement, using a problem-posing instrument to measure one aspect of mathematical creativity (Jensen 1973). This connection between mathematical creativity and giftedness identification in relation to problem posing has also been studied more recently (e.g. Singer et al. 2015). Other studies revealed that problem posing may stimulate creativity, possibly even more than problem solving (e.g. Voica and Singer 2013). In addition, Voica and Singer (2014) found three characteristics that can offer an indication of mathematical giftedness in problem posing contexts: a thorough understanding of conveyed concepts, an ability to generalize reasoning, and a capacity to frame and reframe content in order to devise new problems.

For years, creativity has been studied using four related components outlined by Torrance: fluency, flexibility, originality, and elaboration. Starting from here, various frameworks for studying creativity in relation to giftedness and high achievement have been generated, usually adapted to specific types of tasks. In a problem-solving context, Leikin (2009, 2013) uses multiple-solution tasks as a lens to observe creativity. The dimensions used in her model are originality, fluency and flexibility; these were aggregated into a creativity score by a research-based and, subsequently refined, scoring technique. Leikin and Kloss (2011) examined students' problem solving performance on Multiple Solution Tasks (MSTs) and demonstrated that correctness in problem solving is highly correlated with fluency and flexibility, whereas originality is shown as a special mental quality.

A different approach to creativity, one based on organizational theory, has been taken by Singer and her research team (e.g. Singer and Voica 2013; Pelczer et al. 2013; Voica and Singer 2013). Their framework relies on the concept of cognitive flexibility. Cognitive flexibility is described by cognitive variety, cognitive novelty, and changes in cognitive framing. Cognitive variety manifests in the formulation of different new problems/properties from an input stimulus; cognitive novelty captures the innovative aspect in the posed problem—its distance from the starting element; while changes in the participant's mental frame refer to shifts in the "on-focus" elements during the problem posing. Thus, cognitive flexibility arises as a complex, non-linear interplay between these dimensions. Consequently, the construct of cognitive flexibility opens up the possibility to capture different ways of being creative, namely through the differing loads on the three dimensions. The use of the cognitive-flexibility framework in analyzing data offers more possibilities to capture implications of a social-communicative nature.

By putting fourth to sixth graders, and also university students in problem-posing contexts, Singer and Voica found that, in problem posing situations, high achievers, including gifted mathematics students, develop cognitive frames that make them cautious in changing the parameters of their new problems, even when they make interesting generalizations. The students' capacity to generate coherent and consistent problems in the context of problem modification may indicate the existence of a generalization strategy that seems to be specific to mathematical creativity, differentiating it from creative manifestations in other domains (Singer and Voica 2015). The domain-specificity of mathematical creativity was also identified in other studies (e.g. Kattou et al. 2015). They investigated whether creativity is domain-general or domain-specific by relating fourth through sixth graders' performance on two tests: the Creative Thinking Test and the Mathematical Creativity Test. Their data analysis from 476 students converged on the conclusion that creativity is domain-specific. Other studies found that creativity is not only domain-specific, but it even seems to be task specific within content areas (e.g. Baer 2012).

New studies on the relationships between creativity and giftedness extend their area of research from students' cognitive dimensions to attitudes and values, based on the anticipated roles that individuals with high potential play in society. As some researchers underlined, the biggest challenge in gifted education is to extend the traditional investment in the production of intellectual capital to include an equal investment in social capital, innovation and the development of leadership capabilities (e.g. NSB 2010; Renzulli 2012). The goal should be not simply to ensure that mathematically gifted students fulfill their potential by becoming productive pure and applied mathematicians, but also to ensure that mathematical creativity is enhanced to prepare innovative, thoughtful leaders in all fields with their new, atypical methods and insights.

4   Research into Practice: Programs and Pedagogy

As important as it is to define and recognize students with mathematical gifts, talents, and promise, it is even more important to develop, support and enhance those traits. In this section, we look at examples of teaching practices, tasks, curricula, and in-school and extracurricular programs that are based on the research described earlier and are designed not only to develop mathematical talent and creativity but also to increase mathematical passions by engaging students in problem solving, problem posing, and innovation.

4.1   How Might Teaching Practices Affect Mathematical Promise and Talents?

As noted in Sect. 3, brain plasticity—the ability of the brain to grow and change with learning and experience—has been well documented. Jensen (2000) even stated: "We now know that the human brain actually maintains an amazing plasticity throughout life. We can literally grow new neural connections with stimulation, even as we age. This fact means nearly any learner can increase their intelligence, without limits, using proper enrichment" (p. 149). We may not know whether or not there are limits on how much students might increase their mathematical expertise or just how much a growth mindset as described by Dweck (2006) and others might help, but we do know that teaching practices can greatly increase a student's mathematical performance and passion.

In the United States, the Common Core State Standards for Mathematics (NGA/CCSSO 2010) list eight Standards for Mathematical Practice that have been shown to be effective in developing all students mathematical talents. These standards for students are: Make sense of problems and persevere in solving them; Reason abstractly and quantitatively; Construct viable arguments and critique the reasoning of others; Model with mathematics; Use appropriate tools strategically; Attend to precision; Look for and make use of structure; Look for and express regularity in repeated reasoning. All students should be engaged in these practices throughout their mathematical education, but the CCSSM make no mention of special provisions for gifted, talented, promising or high-achieving students. Noting this oversight, in a joint publication from the NCTM, the National Association for Gifted Children (NAGC), and the National Council of Supervisors of Mathematics (NCSM), titled *Using the Common Core State Standards for Mathematics with Gifted and Advanced Learners,* the addition of a ninth Standard for Mathematical Practice is suggested. "In order to support mathematically advanced students and to develop students who have the expertise, perseverance, creativity and willingness to take risks and recover from failure, which is necessary for them to become mathematics innovators, we propose that a ninth Standard for Mathematical Practice be added for the development of promising mathematics students—a standard on mathematical creativity and innovation: Solve problems in novel ways and pose new mathematical questions of interest to investigate. The characteristics of the new proposed standard would be that students are encouraged and supported in taking risks, embracing challenge, solving problems in a variety of ways, posing new mathematical questions of interest to investigate, and being passionate about mathematical investigations." (Johnsen and Sheffield 2012, pp. 15–16).

To implement these effective practices, Sheffield (2009) recommends that teachers pose problems that allow all students, including the most talented, to struggle; expect coherent explanations and critiques of unique and creative solutions; give formative and summative assessments that provide opportunities for students to reason, create problems, generalize patterns, solve problems in unique ways, and connect various aspects of mathematics; and generally act as a role model who is comfortable with making mistakes and demonstrating the joy of solving difficult problems.

4.1.1 Problem Solving and Problem Posing

Problem solving is often cited as a major goal in any mathematics program. This statement from *Adding It Up*, is typical: "We see problem solving as central to school mathematics. Problem solving should be the site in which all of the strands of mathematics proficiency converge. It should provide opportunities for students to weave together the strands of proficiency and for teachers to assess students' performance on all of the strands" (National Research Council 2001, p. 421). Making sense of problems and persevering in their solutions is the first of the CCSS Standards for Mathematical Practice.

Problem solving is often defined as seeking a solution to a mathematical situation for which students have no immediately obvious process or method. For gifted mathematics students, that means that a question that may be a problem for other students may not engender difficulty for the gifted student. It is important, therefore, that students not only learn to solve problems, but also to rephrase and pose new questions that are authentic problems for themselves, challenging them to persevere and struggle to find a solution.

Several studies on problem solving and problem posing have shown the efficacy of this approach in the development of mathematical creativity and talent. For example, in a 4-year longitudinal study with primary students, Singer found that a pattern of training organized under the name of dynamic structural learning could substantially raise students' creative approaches in mathematical problem solving and posing (Singer 2007). The dynamic structural learning is based on distributing the training procedures across several categories including *systematic training of transfer* (e.g. transfer from objects to various unconventional ad-hoc notations, then to conventional representations, and then to abstract reasoning and back; from thinking aloud to "thinking in mind" and vice-versa; etc.), *randomized training* of the developed capacities, which is realized by means of various mental games, and *structured training* of specific competencies, which aims at assimilating the invariants by constantly resorting to models and diagrams (Singer 2007). In addition, an effective context for both broadening and deepening student's knowledge can be offered by problem posing sessions. The problem-posing research field is an emerging force within mathematics education, which offers a variety of contexts for studying and developing abilities in mathematically promising students (Singer et al. 2013).

4.1.2 Discourse and Questioning

When solving problems, students should discuss their processes, justifying their reasoning, and critiquing their own and their peers methods and solutions. Chapin and colleagues designed and researched what they labeled "talk moves" as part of Project *Challenge*, a Jacob K. Javits grant program of the United States Department of Education that was looking for projects to increase the number of ethnic and linguistic minority students in programs for gifted and talented students. Not only did they find that this oral discourse paid off with more complex, sophisticated, and mathematical reasoning, but students moved from 4 % being classified as "Superior" or "Very Superior" on the Test of Mathematical Abilities, Second Edition (TOMA) at the beginning of the program to 41 % being classified at this level after 2 years in the program. They hypothesized that the discussions allowed misconceptions to surface and be corrected, developed students' ability to reason, gave more students the opportunity to observe, model, build on and add to the development of complex ideas, and provided motivation and engagement (Chapin et al. 2009).

To instigate a rich discussion, teachers and students themselves need questions that assist students in focusing on the big ideas in the problems rather than funneling students to rotely follow a fixed procedure. Sheffield (2006) suggests the use of "who, what, when, where, why and how" questions that are commonly used to teach students to write informational articles. These questions include "*Who* used a different method or has a different solution? *Who* has a new or unique question or suggestion? *What* generalizations or conjectures might I make from the patterns? *What* proof do I have? *What if* I change one or more parts of the problem? *Why* does that work? If it does not work, *why not*? *How* does this compare to other problems or patterns that I have seen? *How many ways* might I use to represent, simulate, model, or visualize these ideas?"

These questions are an integral part of *Project M³: Mentoring Mathematical Minds*, *Project M²: Mentoring Young Mathematicians*, and *Math Innovations* curricula, which will be discussed later in this review, which have been shown to be successful in developing mathematical promise in students with various initial levels of ability.

4.2 How Might Curriculum Contribute to Mathematical Development?

In any mathematics program, in addition to the teaching practices, the curricula and the problems that students encounter often determine whether students have the opportunity to develop their mathematical expertise to the fullest extent possible. One important aspect is the opportunity for students to deepen their understanding of mathematics. *Deepening* is usually associated with studying a curricular topic at greater depth or with greater complexity than prescribed by the curriculum or school textbooks. Deepening mathematical understanding can include, for example, justifying or proving the reasons behind arithmetic operations, solving problems in a variety of ways, or posing and solving related problems. Students might also work on fields of problems. This approach supports competencies via problem solving with a high level of complexity.

4.2.1 Challenging Mathematical Tasks

A common approach around the world to support students in deepening their mathematical understanding is the use of challenging or rich tasks for the realization of mathematical potential. Whether these are called rich tasks, problems, investigations or challenges, these are interesting and motivating mathematical difficulties that a person can overcome. In the publication of the sixteenth International Commission on Mathematical Instruction (ICMI) Study on Challenging Mathematics, Barbeau defines a challenge as "a question posed deliberately to entice its recipients to attempt a resolution, while at the same time stretching their understanding and knowledge of some topic" (Barbeau and Taylor 2009, p. 5). Stretching understanding, making continuous progress in the face of difficulties, and creating new knowledge is especially important for gifted students who are too often faced with repetitive tasks, memorized algorithms or arithmetic skills that they have already mastered. It is also important for challenges to engage the mind, encourage students to explore the beauty and structure of mathematics, and exult in the mastery of new ideas.

Many authors recognize the centrality of mathematical challenge for the realization of mathematical promise and as a characteristic of the activities in which gifted mathematicians are involved. Solutions of challenging tasks often involve explanations, multiple strategies, models and tools, questioning, conjecturing, and ongoing evaluation. Challenges might be multiple-solution tasks, proof tasks, concept-defining tasks, inquiry-based tasks, or other complex tasks that engage students in new mathematical explorations. Mathematical challenge depends on the type and conceptual characteristics of the task, for example, conceptual density, mathematical connections, the building of logical relationships, or the balance between known and unknown elements (e.g. Leikin 2011).

Rich learning tasks are not rich on their own. It depends on what is done with them. Instead of using the straightforward four-step heuristic that is common in many U.S. mathematics textbooks (1. Understand the problem; 2. Devise a plan; 3. Carry out the plan; and 4. Check), Sheffield (2003) has developed a model for a more open heuristic for solving and posing problems where one may start at any point and proceed in any order that makes sense, repeating steps as necessary as they become engrossed in problem solving and problem posing. Steps include relating the task to what students already know; investigating the problem; evaluating the findings; communicating the results; and creating new questions to explore. This heuristic encourages multiple solutions, models and methods as well as problem posing that have been shown to be successful in developing students' mathematical promise.

4.2.2 Curriculum and Textbooks

A mathematics curriculum broadly may be thought of as the total of all the students' mathematical learning experiences. Curriculum may include a body of knowledge to be transmitted as well as the process by which this happens. Thus curriculum writers generally attempt to write materials for teachers explaining expectations for teaching practices as well as student books with specific mathematics content.

Most mathematics textbooks and units are written for a broad range of students. One example of an exception to this is *Project M³: Mentoring Mathematical Minds* (www.projectm3.org) with units that were developed to nurture mathematical talent in elementary students, utilizing the "talk moves", questioning strategies, an open heuristic and challenging tasks as described earlier. With effect sizes ranging from 0.69 to 0.97 on the Open-Response Assessments, results indicated that these units, designed to address the needs of mathematically promising students, positively affected their achievement (Gavin et al. 2009). Following the success of *Project M³*, *Project M²: Mentoring Young Mathematicians* (www.projectm2.org) was developed with support from a grant from the US National Science Foundation. Units in this program were designed for heterogeneous classes of students from kindergarten through second grade. One purpose of the M^2 program was to determine whether using the same "talk moves", questioning strategies, open heuristic and challenging tasks described earlier could increase the numbers and levels of mathematically talented students. Following participation in the program, results showed a significant difference at the 0.001 level in favor of M^2 students in the percent of students performing one and two standard deviations above the mean between students in the M^2 program and students in the comparison groups even though groups were not significantly different before the beginning of the program (Gavin et al. 2009, 2013; Sheffield et al. 2012).

4.3 What In-School Programs Might Develop Mathematical Talent?

Countries vary widely in programs to identify, support, create and enhance students with mathematical expertise and passion. A brief overview of the status of gifted education across the world is provided in this section. The World Council for Gifted and Talented Children (WCGTC/ www.world-gifted.org) provides worldwide advocacy and support, has affiliated federations in Africa, Asia-Pacific, Europe, and Ibero-America with organizations and resources specific to those areas, and holds international conferences every two years. The United States has no federal policy on gifted education, but The National Association for Gifted Children (NAGC) www.nagc.org/ regularly surveys gifted programs across the US, and posts information on their website. In Africa, Mhlolo (2014) reported on a survey of 15 African countries to determine the extent to which mathematically talented students were identified, tracked and nurtured, and in Europe, Mönks and Pflüger (2005) surveyed 21 European countries concerning legislation, identification, provisions, teacher training, research and priorities in gifted education. A few of the in-school programs and activities specific to the support and enhancement of mathematical promise and talents are described here.

4.3.1 Ability Grouping, Self-contained Classes and Specialized Schools

In a survey of over 1000 school districts in the United States, Callahan et al. (2014) reported that while over 90 % of the districts claimed to identify gifted students, services for these students varied. About half the elementary programs responding said that they had special homogeneous classes for gifted students pulling them from heterogeneous classes from 1 to 4 h a week, about two-thirds of the middle schools reported the existence of some special homogeneous classes and 90 % of the high schools reported using Advanced Placement® as the predominant option. Mönks and Pflüger (2005) found that 12 of the 21 European countries surveyed reported that giftedness (or a synonym such as high-ability, or talented) was explicitly named in the law of the country, 13 countries reported differentiation for these students, and 15 reported having a special curriculum. In a survey of 15 sub-Saharan African countries, Mhlolo (2014) found that the most common method of identifying mathematically talented students was through participation in a mathematics Olympiad in 12 of the 15 countries with no method of identification named in the others. Many mentioned that the Olympiads were used to recognize mathematical achievement but nothing was done to support or nurture it. Gifted students were placed in regular classrooms per national policies of non-streaming of students. No country mentioned having special public schools for gifted students or providing any special training for teachers of the gifted.

One method of serving mathematically talented students around the world is through ability-grouping, self-contained classes and specialized secondary schools for the mathematically talented where students are tested and must show some level of general giftedness and/or mathematical expertise before being accepted into the program. One issue that has triggered many debates all-over the world is *ability grouping classroom versus heterogeneous classroom.* The debate on the necessity of ability grouping is legitimate, and both proponents and opponents of heterogeneous mathematics education use valid arguments to justify their positions (Leikin 2011).

Some studies suggest that ability grouping may be essential for the education of gifted both from cognitive and affective perspectives and that therefore, schools ought to supply special education to mathematically gifted students and prevent talent loss (Milgram and Hong 2009). Olszewski-Kubilius (2013), past President of the National Association for Gifted Children in the US, reports that a 2013 National Bureau of Economic Research study of students who were grouped by ability found that the performance of both high and low performing students significantly improved in mathematics and reading. On the other hand, ability grouping is still questionable both in light of the equity principle and of some research findings. Some of this may be due to different definitions of ability grouping. In the United States, tracking which identifies students at a given point in time and places them on a semi-permanent, rigid, defined path is distinguished from flexible grouping where students are not given permanent levels and may move up or down based on performance.

At the high school level, several countries have special schools for talented mathematics students. In *Special Secondary Schools for the Mathematically Talented: An International Panorama,* over 100 special schools from twenty countries are described ranging from the Super Science High Schools in Japan and the Science High Schools and Gifted High Schools in Korea to the special schools in Europe, Russia, North and South America, the Middle East, China, South Asia, and Australia. These schools have shown to be effective in offering exciting frameworks for the education of gifted students (e.g. Vogeli 2016; Vogeli and Karp 2003). Many of today's leading mathematicians and mathematics educators have come and will continue to come from these specialized secondary schools for the mathematically talented (Vogeli 2016). In the United States, the National Consortium of Secondary STEM Schools (www.ncsss.org) is designed to allow these schools to share and build on each other's successes and challenges. The new volume from Vogeli (2016) facilitates this sharing on a worldwide level.

4.3.2   Acceleration and Grade Skipping

Acceleration is usually defined as learning topics within the curriculum at a faster pace. In a comprehensive study of acceleration that includes 13 different types of acceleration among which grade-skipping, moving ahead in one subject area, advanced placement, curriculum compacting, dual enrollment in high school and college classes, and entering college early, Colangelo et al. (2004, p. xi) found that "acceleration is educationally effective, inexpensive, and can help level the playing field between students from rich schools and poor schools". Even for moderately gifted students, recent research shows that approximately 40–50 % of traditional classroom material could be eliminated for targeted gifted students in one or more of content areas, among which is mathematics (Reis et al. 1998). Care must be taken not to skip critical material, however, and to ensure that students are engaged and passionate about the mathematics they are learning, and are not simply rotely memorizing algorithms or accelerating so they can finish taking required mathematics classes early.

A caveat must be added as well for students whose mathematics program has been accelerated by simply moving faster through a traditional program. In the United States where the numbers of STEM majors in college decreased between 1984 and 2010 while the number of students taking calculus in high school skyrocketed, Bressoud et al. (2012, p. 2) note that "What the members of the mathematical community—especially those in the Mathematical Association of America (MAA) and the National Council of Teachers of Mathematics (NCTM)—have known for a long time is that the pump that is pushing more students into more advanced mathematics ever earlier is not just ineffective: It is counter-productive. Too many students are moving too fast through preliminary courses so that they can get calculus onto their high school transcripts. The result is that even if they are able to pass high school calculus, they have established an inadequate foundation on which to build the mathematical knowledge required for a STEM career."

4.4   What Extra-Curricular Programs Might Enhance Mathematical Promise?

Programs for gifted mathematics students that are offered during the school day are often supplemented and enhanced by extra-curricular programs. At other times, extra-curricular programs are the only way that mathematical expertise is identified, challenged and enhanced.

4.4.1 Recreational Mathematics

As noted earlier, pullout programs for a few hours a week for identified gifted elementary students in the United States who spend the majority of time in heterogeneous classes are a common service option as is differentiation or cluster grouping within a heterogeneous class. Other activities such as math clubs, competitions, online courses, project learning, and work with mentors can be found both in school and out of school. In all of these settings, teaching practices and challenging tasks are important for the fostering of mathematical talent.

Brandl (2014) analyzed different settings of fostering mathematical talent with respect to the students' attitude towards mathematics. The findings suggest that, on the one hand, the environment of an ordinary mathematics class often seems not to be supportive enough for promoting mathematics as something beautiful, challenging and joy-bringing; on the other hand, selection with respect only to performance and high achievement can lead to a reported "psychological hindrance of a *narcissistic wound/shock* that comes from being confronted with just best-of-students in class and the eventual loss of this status for oneself" (Brandl 2014, p. 1164). So, non-selective interest-based courses seem to be more promising.

A variety of extra-curricular options exist to engage students in mathematical explorations. Many of these focus on the enjoyment of recreational mathematics. In many parts of the world, students routinely turn to reading both fiction and non-fiction as an enjoyable pastime, but "doing math" for fun has not been as popular. These activities exist to give students an enjoyable mathematical option.

The term math "club" is often used to describe any extra-curricula mathematics program designed as a fun way to challenge students to encounter interesting mathematics. Math circles are generally programs where college and university mathematicians share their expertise and love of mathematics with K-12 students and teachers. They follow a variety of styles from informal activities and games to more traditional enrichment classes. Some have a strong emphasis on preparing students for competitions while others avoid all competitions. Math Circles appeared in Russia in 1930 and have existed in Bulgaria and Romania for over a century. Math circles migrated to the United States in the 1990s, often with immigrants who enjoyed these activities themselves as teenagers, and have grown in popularity in the US in the last 25 years. The National Association of Math Circles (NAMC, www.mathcircles.org) provides resources and support for Math Circles and other similar informal mathematics education programs around the world. Math houses (www.mathhouse.org) in Iran serve a similar function to the math circles with activities and competitions for students and workshops and meetings for their teachers from the primary through the university level.

There are numerous resources for recreational math (for example, see lists from the Mathematics Association of America (www.maa.org/programs/students/student-resources) and the NCTM (www.nctm.org/Classroom-Resources/Browse-All/), including videos and websites with challenging problems, games, and other interactive resources. There are also extended extra-curricular programs for engaging mathematics such as the summer and week-end math camps that are listed by the American Mathematical Society (www.ams.org/programs/students/emp-mathcamps), or the activities and camps related to the Kangaroo contests (e.g. www.mathkangaroo.org; mathplus.math.toronto.edu/).

Online enrichment problems such as those at CAMI (www.umoncton.ca/cami), a website (in French) developed to provide all students with challenging and rich mathematical problems online following a Problem-of-the-Week model to which all students could submit their solutions and receive a feedback from mentors—pre-service teachers from a local university, are another way to challenge and engage students. Results from the use of this site with over a million hits from more than 100,000 visitors between 2005 and 2010 found that students seemed to appreciate problems that were different from those they encountered in the regular classroom and solving them was more challenging, motivating and enjoyable, even if sometimes problems seemed to be too difficult and frustrating. Personal feedback from mentors was also greatly appreciated (Freiman 2009; Freiman et al. 2009; Freiman and Lirette-Pitre 2009; Freiman and Manuel 2015).

4.4.2 Competitions

There are a variety of math competitions around the world. The most known of these is the International Mathematical Olympiad (www.imo-official.org), which is the World Championship Mathematics Competition for High School students and is held annually in a different country. The first IMO was held in 1959 in Romania, with 7 countries and has gradually expanded to over 100 countries from 5 continents. Other international math competitions can be found at www.artofproblemsolving.com/. Preparing students for this type of competition is somewhat different from some of the programs based on problem solving or recreational mathematics. With these competitions it is often necessary to acquire mathematical knowledge, and learn algorithms, theorems, and mathematical "tricks" explicitly. The difference lies in the speed of working on the questions. Students are successful if they "see" immediately the mathematical core of the question and if they can embed the question in the mathematical background. This is different than other types of problem solving or problem posing where the students have time for going a long way around, for trying different approaches and by this acquire more and more knowledge about the subject as well as about metacognitive aspects.

Not all competitions are of that same type, however. There are game competitions such as: Set (www.setgame.com); Math Pentathlon (www.mathpentath.org); Kangaroo, which is the largest mathematical competition in the world, with more than six million participants from 72 countries in 2015 (http://www.aksf.org/); or Calculation Nation (calculationnation.nctm.org/), puzzles like Ken-Ken (www.kenken.com) and competitions that give students an extended period of time to solve interesting problems such as the USA Mathematical Talent Search (www.usamts.org) that gives students at least a month to work out problems with written explanations and encourages the use of any materials including books, calculators, and computers. Like other programs and activities in this section, regardless of the type of competition, they are designed to encourage and reward high-level mathematical performance, and hopefully to entice more students to continue with mathematics-intensive studies and careers.

Despite the variety of programs for the education of mathematically gifted students, there is lack of systematic empirical studies on various programs to gain better understanding of their suitability for the realization of students' mathematical potential. Theoretical characterizations of effective courses and programs for mathematically talented students, as well as empirical studies to test their effectiveness are still to be developed. More information is needed on just how extensive, widespread and effective are programs that are designed to serve or develop students with mathematical promise or expertise. Unfortunately, in many parts of the world, no such programs exist, and in many cases, the programs that do exist are small, insignificant or not optimally effective. This results in untold loss of needed mathematical knowledge and aptitude in citizens and educators around the world, in addition to the loss of expertise in science, technology, engineering and many other math-intensive fields.

5   Research into Practice: Teacher Education

Teachers, of course, are critical to implanting the programs and practices that have been shown to be effective with mathematically promising students. In addition to traits and professional development that are important for all teachers of gifted students, there are supplementary expectations from teachers of mathematically gifted students.

5.1   What Teacher's Traits Are Important?

Greenes et al. (2010) noted that research findings indicated that teachers of gifted and talented should be flexible thinkers who are appreciative of creative approaches, curious, persistent, and confident when solving difficult problems. In addition, for teachers of mathematically talented students, she also recommended that they should understand a wide range of mathematics concepts and skills, have a "toolbox" of problem solving heuristics, love math, and be armed with challenging and engaging mathematical problems. Chamberlin and Chamberlin (2010) summarized the needed competencies of teachers of gifted students as knowledge of student needs; skill in promoting higher-level thinking, creativity and problem solving; development of a differentiated curriculum with multiple resources, enrichment and acceleration; creation of a safe, flexible, learner-centered environment; and avoidance of rote memorization and overreliance on gifted students as tutors.

Leikin (2011) used one example of a teacher of mathematically gifted she called 'outstanding' to reveal several similar characteristics that his students valued the most, with the addition of a category on the teacher's relationship to the students. Among the characteristics that students valued in a teacher, one category represented relationship to the subject, namely, genuine interest, deep knowledge, including knowledge that went beyond mathematics, and openness to challenge from the students. The second category featured relationship to the students, including kindness, trustworthiness and pride along with patience and sensitivity to students' interests, needs, difficulties and differences. The third category referred to teaching style and methods such as being flexible, knowledgeable, creative, and open to improvisation, as well as having a sense of humor, love of and enjoyment from mathematics.

According to Holton et al. (2009), the teacher's role is central in promoting the mathematical understanding and learning of students by choosing appropriate tasks and providing expert assistance. First, the teachers must be able recognize the importance of balancing teaching mathematical rules and algorithms, on the one side, and more complex and sophisticated mathematical processes and open problems on the other side, as part of a more creative work. Second, they need to be aware of students' cognitive and social processes including a theoretical knowledge of how students learn. Third, be able to adjust teaching to the result of interaction with the students and to encourage both oral and written communication among students. Finally, they need to be aware of the nature and importance of mathematical challenge.

Several researchers have noted difficulties that teachers have in challenging promising mathematics students. For instance, regarding the use of challenging tasks by teachers, Holton et al. (2009) point at different barriers, such as the lack of attention to the process side in students' mathematics learning, and particularly, in supporting cooperative processes of discovery learning or teamwork, and also low expectation-levels from what are the students' abilities (p. 216). Along with the lack of teachers' own motivation to do challenging tasks, other, more systemic obstacles related to the social and educational policies and economic conditions can also affect teachers' willingness to use challenging tasks with their students.

Leikin (2011) found that simply providing teachers with challenging mathematics activities is not sufficient for their implementation. She noted that teachers have to be provided with multiple opportunities to advance their knowledge and to develop commitment and beliefs in their own and students' abilities for high-level mathematical performance. Consequently, she lists the following questions that need additional research:

- Should the teachers of gifted be gifted? Should the teachers be creative in order to develop students' creativity?
- How might teachers' creativity be characterized both from the mathematical and from the pedagogical points of view?
- What are the desirable qualities of teachers' knowledge, beliefs and personality that make them creative and gifted teachers?

Answers to these questions may lead to more effective teaching, which is oriented towards helping students to reach their cognitive potential and to develop a balanced personality. This leads us to the next question.

5.2 What Should Be Included in Teacher Education?

The recommendations in this section for teachers' preparation and professional development are based on the proposed teacher's traits and knowledge described in the previous section.

5.2.1 Teachers of All Students

In the United States, the National Association for Gifted Children (NAGC) and the Council for Exceptional Children (CEC) collaborated to develop the *NAGC-CEC Teacher Preparation Standards in Gifted Education* to assist state departments of education in developing standards for teachers of K-12 gifted students and to help colleges and universities develop pre-service and in-service teacher education programs to prepare and support teachers of gifted students (NAGC-CEC 2013). These standards were followed by the *Knowledge and Skill Standards in Gifted Education for All Teachers.* Those standards stated that all K-12 teachers should be able to recognize learning differences, cognitive/affective characteristics and needs of gifted and talented students of all backgrounds and design appropriate learning modifications to enhance creativity, acceleration, as well as depth and complexity of learning, by using a repertoire of instructional strategies to advance their learning (NAGC-CEC 2014).

Tomlinson (1999) analyzed differentiation in the classroom and pre-service teacher preparation, viewing it as a way to meet the needs of gifted and other academically diverse students. In the conducted experimental studies, Tomlinson and colleagues found that although participants affirmed the existence and importance of recognizing student differences and concomitant needs, they used ambiguous criteria for identifying these differences and needs, expressed incomplete views of differentiating instruction, exhibited shallow wells of strategies for enacting differentiation, and were influenced by factors which complicated and discouraged understanding and addressing student differences and needs. Presenting these teachers with workshops on strategies for more differentiated instruction and coaching them when implementing these strategies in schools may be a promising way of better teacher preparation; yet, several issues arose when translating training into the real practice due, for example, to the lack of collaboration between the coach, the teacher and the university supervisor.

This risk was addressed by Singer and Sarivan (2009), who developed a strategy of preparing teachers called *multirepresentational training* (MRT) that has been applied in graduate and undergraduate mathematics courses for primary prospective teachers. The MRT model is based on two directions of action: providing a variety of representations as powerful tools for learning, and developing representational models to stimulate abstraction and synthesis (Singer 2007, 2009). The contextualized learning paths developed through the MRT help students-prospective teachers reach a depth of understanding that enables them to reiterate their learning acquisition in the different and complex problem-solving contexts of teaching in the real setting (Singer and Sarivan 2009, 2011). This method worked as a shortcut for internalizing mathematical content knowledge, mathematics pedagogical content knowledge, and general pedagogical knowledge in an integrated effective way.

5.2.2   Teachers of Mathematically Gifted and Talented Students

The NCTM Task Force Report on teaching mathematically promising students suggest including information on dealing with mathematically promising students in in-service and pre-service programs for teachers at all levels. Regardless of the type of program being offered to promising students, teachers should have access to professional development, research information, and resources to deal with such issues as identification or recognition of students with mathematical promise, high levels of expectations for all students along with challenging top students to even higher levels of success, pedagogical and questioning techniques to extend students' thinking, and selection and/or development of appropriate curriculum and assessment tools that provide opportunities for students to create problems, generalize patterns, and connect various aspects of mathematics, development of teachers' own mathematical power to make connections and the mathematical sophistication to see the big picture, making appropriate instructional decisions for these promising students, and awareness of, access to and ability to use technology and other tools. In addition, teachers should continue to strengthen their own mathematical content knowledge and demonstrate the joy of being a lifelong learner of mathematics (Sheffield et al. 1999).

Holton et al. (2009) note the importance of modeling the incorporation of challenging tasks programs for teachers asking them to construct meaning and make explicit connections among mathematical ideas and to prior knowledge. Leikin and Winicky-Landman (2001) note that both teachers' Pedagogical Content Knowledge and Mathematical Content Knowledge are enhanced when the teachers cope with mathematical challenges as learners. Greenes and Mode (1999) also note the importance of teachers of mathematically gifted strengthening their own mathematical content knowledge through individual, partner, and small-group problem solving during pre-service courses and in-service workshops. They suggest that prospective teachers should individually assess and mentor students who have been identified as mathematically promising and develop individualized learning plans for them that identify specific areas of mathematical talent, interests and needs; set goals to be achieved; and identify resources, challenging problems, and strategies for accomplishing the goals as well as assessments to monitor progress. They also recommend that mathematics teachers observe and plan with each other; analyze, critique, and maintain a file of students' most outstanding mathematics work; tackle challenging mathematics problems themselves; identify and provide engaging extracurricular mathematics activities; join professional societies and seek out opportunities for ongoing professional development; and learn strategies for academic and career counseling.

For teachers of promising secondary mathematics students, Greenes et al. (2010) specifically recommend a sequence of five workshops or courses that go beyond their initial teaching certification. These courses include a problem-solving lab; classes on assessing mathematical talent and differentiating instruction; instructional strategies for this category of students; and a concept study focusing on a variety of ways in which students might develop deep understanding of a key mathematical concept.

More research is needed to validate different strategies for enhancing the quality of professional development of the teachers of the gifted, and specifically on the ways in which such strategies can be embedded into the initial preparation for the teaching profession of all (future) teachers.

5.3   What Are Some Examples of Programs for Supporting Teachers?

As mentioned earlier, Mhlolo (2014) in his survey of Sub-Saharan African countries found none that offered teacher training specifically for teachers of gifted and talented students, but Mönks and Pflüger (2005) in their survey of European countries found 16 of the 21 countries offered some type of teacher training on gifted education. In the United States, the *NAGC-CEC Teacher Preparation Standards in Gifted and Talented Education* (2013) are included in accreditation standards for colleges and universities that choose to have their graduate programs for preparing teachers of the gifted accredited through the Council for the Accreditation of Educator Preparation (CAEP). Callahan et al. (2014) found that the amount of professional development that in-service teachers received related to gifted and talented students varied widely from district to district, ranging from 0–15 min to 4 days per year. However, those surveys refer to general gifted education professional development that is not necessarily focused on mathematics. The present section includes a few examples of professional development programs for teachers of mathematically promising students in different countries.

5.3.1   Brief Snapshots into Professional Development Programs

A pilot study of a professional-development course for science and mathematics teachers offered as part of a 2-year in-service program for promoting excellence in education was conducted in Israel by Karsenty and Friedlander (2008). The course aimed to expose teachers to theoretical aspects of gifted education in general, and particularly in science and mathematics; to develop leadership qualities based on classroom contexts provided by teachers; and to increase teachers' domain-specific pedagogical content knowledge. Five types of activities were developed as answers to the above questions: (1) analysis of lessons and interviews with gifted students in order to learn about their cognitive and affective characteristics; (2) analysis of investigative activity (launch-explore-summarize) from the students' point of view; (3) analysis of mathematical tasks for advanced students (mathematical content, context, level of openness, representations and sequence of sub-tasks); (4) design, adaptation or adoption of tasks: learning about strategies for adapting routine mathematical tasks to the needs of the gifted (e.g. What if not? problem-posing strategy); and (5) classroom implementation of activities that are then reported and analyzed (Karsenty and Friedlander 2008). These activities have been shown to have a strong potential for impact on prospective teachers.

A large gamut of studies addresses the issues of professional development for teachers of mathematically promising students in the United States. For example, Adelson et al. (2007) discussed teachers' professional development for specific programs, like *Project M^3: Mentoring Mathematical Minds*, which was based on an enriched and accelerated curriculum focused on developing conceptual understanding in mathematics (see also Sect. 4.2.2). The teachers involved in this project participated in a two-week summer training program in order to increase their mathematical content knowledge and to learn how to implement teaching strategies to promote deeper reasoning, problem solving and problem posing, and verbal and written mathematical communication. Teachers also attended four to six professional development sessions throughout the academic year prior to teaching each unit of the curriculum. A professional development team member visited each school every week the *Project M^3* units were taught, to ensure fidelity of treatment and offer individualized assistance to teachers. Students made highly significant progress, while teachers reported that this embedded professional development contributed to strengthen their own mathematical content knowledge and understanding of their students, and led to an enhancement of their repertoire of teaching practices and strategies.

In a study by Chamberlin and Chamberlin (2010), pre-service teachers enrolled in a mathematics teaching methods course, without having special training in teaching the gifted, were asked to choose tasks they found appropriate to use with gifted students and to implement them in a real classroom setting. According to participants' reports, teachers seemed to have broadened their view of giftedness, recognized the need to adapt instruction for gifted students, made efforts to align problem-solving tasks with gifted students' readiness and interests, realized the necessity of knowing students to differentiate instruction, and emphasized student-centered approaches.

These studies, as well as several others, point to the need for teachers to have experience with the use of challenging tasks and the appropriate feedback to be given to students.

5.3.2 Professional Development Related to Mentoring Students' Online Problem Solving

Noting that the analysis of children's mathematical production by pre-service teachers has become an important part of mathematics education courses, little is known about the impact of participation of pre-service teachers in online activities with schoolchildren and even less about their capacity to guide young learners by means of asynchronous feedback (LeBlanc and Freiman 2011). In a study in Canada, pre-service teachers served as mentors in a context of a virtual mathematics learning community, CASMI (Freiman and Lirette-Pitre 2009) to assess students solving mathematically rich, contextual, and open-ended problems posted on a website that is meant to challenge all children. This activity aimed to help pre-service teachers appreciate the diversity of solutions and learn how to guide schoolchildren in a personalized and caring manner, nurturing their curiosity, interest and perseverance that are very important for all children and especially for the gifted ones. It was concluded that participation in the online project allowed pre-service teachers to experience new mathematical problem-solving approaches that stress the use of multiple strategies and communication means by schoolchildren. At the same time, in a context of asynchronous assessment with no opportunity to give feedback in another way than written comments, a good understanding of a child's reasoning appears to be not an easy task. Participants overlooked some plausible, even ingenious solutions, alternative views, as well as various misinterpretations and misconceptions. LeBlanc and Freiman (2011) point at the potential of 'feed-forward' pedagogy while stressing the need to reinforce pre-service teachers' own conceptual understanding of mathematics and develop a better understanding of how children think and explain their thinking by practicing more their ability to understand the problem itself for an effective guiding of students.

From a different perspective, but in the same idea of personalizing learning and feedback via on-line tools, the System of Testing, Analyzing and Reporting for students (STARs) is based on collaborative databases of questions (items) and allows assessing students' mathematical competence by using tests from the databases, followed by an individualized feedback, obtained by processing student's answers based on a multi-criteria analysis. Further, the student can receive new sets of questions situated in his/her identified range of proximal development. For this system to be effective, the questions need to capture essential elements of mathematics understanding and mathematics creativity (Singer and Singer 2010). To date, the implementation of this system created opportunities for some teacher professional development studies. In one of the studies, in-service mathematics teachers at secondary school level, involved in Mathematics Olympiads training, participated in a two-week summer institute focused on teachers' ability to pose multiple choice problems that would assess student understanding. Two major aspects were identified. On the one hand, a certain resistance of teachers to shift from the mathematical content of a posed problem toward interpretations of students' thinking in relation to that problem was identified; often, the posed-problem formulation was elliptic or full of ambiguity, while the background topic was irrelevant for students' motivation. On the other hand, a change of participants' behavior happened on two dimensions, both observable in group interactions: an openness to discuss and analyze the quality of their own posed problems, as well as an emerging awareness on the need of conceptual understanding of their students' thinking. The training program, thus, contributed to the development of a reflective attitude toward addressing tasks to students (Pelczer et al. 2014). It seems that an in-service training program that systematically combines group interactions with individual problem-posing tasks exploited during further interactions could significantly influence the building of tasks that are focused on students learning with understanding.

5.3.3   Administrative Changes and Cross-Country Studies

In the former USSR, in special schools for mathematically gifted and talented, mathematics was often taught by professional mathematicians who themselves had attended such type of schools (e.g. Freiman and Volkov 2004; Karp 2016). In Canada, with a provincially governed school system, New Brunswick is developing an inclusive model of schools where students of all abilities must be provided with appropriate learning opportunities within a regular classroom setting; here, a more generalist approach to teacher preparation for such context was adopted (see more in Freiman 2010).

In Israel, *The Division of Gifted Education of Israeli Ministry of Education* encourages teachers to get special education, though there is still a shortage of corresponding programs. Among examples of such initiative, Applebaum et al. (2011) mention special teaching certification programs (in three teacher training colleges and two Universities) and the first M.A. program (in Haifa University) devoted to the education of gifted students. However, these programs are mainly interdisciplinary and are not focused on specific school subjects. Applebaum et al. (2011) investigated prospective teachers' conceptions in these two programs about teaching mathematically talented students in Canada and Israel by addressing the issues of teachers' capacity to solve challenging tasks and their views on mathematics education of mathematically promising students. By proposing a particular open-ended task to the participants from both countries (Canada and Israel) and analyzing responses, along with conducting a survey and a group discussion with participants about their own capacity to deal with the task, the strategies they use to solve the problem, the nature of the task suitable for mathematically talented students, and their own preparations to work with these students, it was found that teachers cope with the challenging task with varying levels of success. The majority used 'non-systematic' strategies, without analysis of the efficiency of the strategies. It was also found that Israeli teachers used both non-systematic strategies and systematic ones (that they have previously learned in a different context), whereas most Canadian prospective teachers used mainly non-systematic strategies. The discussion and questionnaire confirmed that participants in both countries acknowledged the importance of challenging and open-ended tasks, sustaining also the need for a special curriculum for mathematically talented students. However, they themselves did not feel prepared for dealing with such tasks in their classroom. Results imply that teachers need better mathematical preparation in terms of solving open-ended challenging tasks that would enable them not to limit students' problem-solving processes with the finding of one suitable solution. Acquiring such cognitive and meta-cognitive skills will help teachers in guiding their students on the way to deeper and more meaningful mathematical knowledge (Applebaum et al. 2011). At a more general level, it implies that teacher education programs should, first, expose teachers to the complexities of teaching mathematically promising students, which might be beneficial for all students in the math class; second, develop teachers' capacities to investigate challenging tasks by proposing such tasks and explorations during their training; and third, amplify teachers' didactical inventory of teaching strategies to allow identification and fostering of students' abilities using inquiry-based challenging tasks.

6 Summary and Looking Ahead

The discussion about giftedness and intellectual power in children of various ages is ultimately a discussion about future leadership and a precious human capital resource. Consequently, the research in this area has a strong social impact.

Assuming this impact, the present survey started by offering a basic description of the nature of mathematical giftedness, then offered a multidimensional analysis of mathematical promise in students of various ages, covering cognitive, social, and affective aspects, recent research in cognitive science and neuroscience, the relationships between novice knowledge and expertise, as well as the interplay between giftedness and creativity. The next two chapters moved from research into practice, focusing on programs and pedagogy for educating mathematically promising students and respectively on teacher education. The programs and pedagogy under discussion referred to practices that could best encourage mathematical promise and talents, approaching problem solving and problem posing, discourse and questioning, creativity and innovation, challenging mathematical tasks, curriculum and textbooks, in-school programs and activities (with reference to ability grouping, self-contained classes and specialized schools, acceleration and grade skipping), and extra-curricular programs and activities (with reference to recreational mathematics and competitions). Given the social impact discussed above, the initial and continuous professional development of teachers is of a major importance. Our survey recorded features of effective teachers; structure and content of teacher-education for teachers of mathematically gifted and talented students (but also for all teachers because a promising student can be everywhere); and examples of successful programs for preparing and supporting teachers of mathematically gifted students in a variety of country-specific contexts.

Our overview recorded the advancement in the research and practices in this emerging interdisciplinary field of mathematical promise in youth. Still, many questions remain unanswered and they can orient further research to be carried out in the following areas:

- The nature of classroom culture and the role of the teacher in fostering mathematical expertise
- The types of curriculum that support individualization and differentiation of learning
- The use of neuroimaging techniques to inform the learning and teaching of gifted and talented students
- The development and use of digital tools to facilitate personalized effective learning
- The nature of professional development that supports teachers' capacity to foster mathematical promise in as many students as possible
- The impact of various frameworks of giftedness treatment for later professional careers.

Acknowledgments We would like to give special thanks to Professor Dr. Marianne Nolte of the University of Hamburg for her invaluable contributions and assistance in the development of this paper.

References

Adelson, J., Carroll, S., Casa, T., Gavin, M., Sheffield, L., & Spinelli, A. (2007). Project M^3: mentoring mathematical minds—A research-based curriculum for talented elementary students. *Journal of Advanced Academics, 18*(4).

Albon, R., & Jewels, T. (2008). Gifted university students: Last chance to 'come out of the closet'. In *10th Asia-Pacific Conference on Giftedness, Singapore*. Retrieved from http://works.bepress.com/cgi/viewcontent.cgi?article=1000&context=rozz_albon

Anderson, K. L., Casey, M. B., Thompson, W. L., Burrage, M. S., Pezaris, E., & Kosslyn, S. M. (2008). Performance on middle school geometry problems with geometry clues matched to three different cognitive styles. *Mind, Brain, and Education, 2*(4), 188–197.

Applebaum, M., Freiman, V., & Leikin, R. (2011). Prospective teachers' conceptions about teaching mathematically talented students: Comparative examples from Canada and Israel. *The Montana Mathematics Enthusiast, 8*(1–2), 255–290.

Assmus, D. (2016). Characteristics of mathematical giftedness in early primary school age. To appear in the *Proceedings of ICME13*. Hamburg, Germany.

Baer, J. (2012). Domain specificity and the limits of creativity theory. *The Journal of Creative Behavior, 46*(1), 16–29.

Barbeau, E., & Taylor, P. J. (Eds.). (2009). *Challenging mathematics in and beyond the classroom (The 16th ICMI Study)*. New York: Springer.

Bicknell, B. (2008). Who are the mathematically gifted? Student, parent, and teacher perspectives. In *Proceedings of ICME11*. TG6: Activities and Programs for Gifted Students.

Binet, A. (1909). *Les idées modernes sur les enfants*. Paris: Flammarion.

Boaler, J. (2015). *Mathematical mindsets: Unleashing students' potential through creative math, inspiring messages and innovative teaching*. San Francisco, CA: Jossey-Bass.

Brandl, M., & Barthel, C. (2012). A comparative profile of high attaining and gifted students in mathematics. In *ICME-12 Pre-proceedings* (pp. 1429–1438).

Brandl, M. (2011). High attaining versus (highly) gifted pupils in mathematics: a theoretical concept and an empirical survey. In M. Pytlak, E. Swoboda, & T. Rowland (Eds.), *Proceedings of CERME 7* (pp. 1044–1055). Univ. of Rzeszów, Poland.

Brandl, M. (2014). Students' picture of and comparative attitude towards mathematics in different settings of fostering. In B. Ubuz, C. Haser, & M. A. Mariotti (Eds.), *Proceedings of CERME 8* (pp. 1156–1165). Ankara: Middle East Technical Univ.

Bressoud, D., Camp, D., & Teague, D. (2012). *Background to the MAA/NCTM statement on calculus*. Reston, VA: NCTM.

Budak, I. (2012). Mathematical profiles and problem solving abilities of mathematically promising students. *Educational Research and Reviews, 7*(16), 344–350.

Callahan, C. M., Moon, T. R., & Oh, S. (2014). *National surveys of gifted programs: Executive summary*. Charlottesvile: Univ. of Virginia, NRCGT. Retrieved January 24, 2016, from http://www.nagc.org/resources-publications/resources-university-professionals

Chamberlin, M. T., & Chamberlin, S. A. (2010). Enhancing preservice teacher development: Field experiences with gifted students. *Journal for the Education of the Gifted, 33*(3), 381–416.

Chapin, S. H., O'Connor, C., & Anderson, N. C. (2009). *Classroom discussions: Using math talk to help students learn*. Sausalito, CA: Math Solutions.

Cipora, K., Patro, K., & Nuerk, H. C. (2015). Are spatial-numerical associations a cornerstone for arithmetic learning? The lack of genuine correlations suggests so. *Mind, Brain, and Education, 9*(4), 190–206.

Clark, B. (2002). *Growing up gifted: Developing the potential of children at home and at school* (6th ed.). Upper Saddle River, J: Prentice Hall.

Colangelo, N., Assouline, S. G., & Gross, M. U. M. (2004). *A nation deceived: How schools hold back America's brightest students*. Iowa City, Iowa: The C. Belin & J. N. Blank International Center for Gifted Education and Talent Development.

Craft, A. (2005). *Creativity in schools: Tensions and dilemmas*. London: Routledge.

Cross, T. L., Cassady, J. C., Dixon, F. A., & Adams, C. M. (2008). The psychology of gifted adolescents as measured by the MMPI-A. *Gifted Child Quarterly, 52*, 326–339.

Dai, D. Y. (2010). *The nature and nurture of giftedness: A new framework for understanding gifted education*. New York: Teachers College Press.

Dehaene, S., Piazza, M., Pinel, P., & Cohen, L. (2003). Three parietal circuits for number processing. *Cognitive Neuropsychology, 20*, 487–506.

Diezmann, C. M., & Watters, J. J. (2000). Characteristics of young gifted children. *Educating Young Children, 6*(2), 41–42.

Dweck, C. (2006). *Mindset: The new psychology of success*. New York: Random House.

Freehill, M. (1961). *Gifted children*. New York: MacMillan.

Freiman, V. (2009). Mathematical enrichment: Problem-of-the-week model. In R. Leikin, A. Berman, & B. Koichu (Eds.), *Creativity in mathematics and the education of gifted students* (pp. 367–382). Rotterdam: Sense Publishing.

Freiman, V., Kadijevich, D., Kuntz, G., Pozdnyakov, S., & Stedoy, I. (2009). Challenging mathematics beyond the classroom enhanced by technology. In E. Barbeau & P. Taylor (Eds.),

The 16th ICMI study. New ICMI Study Series (Vol. 12, p. 325). *Challenging mathematics in and beyond the classroom*. Springer.

Freiman, V., & Lirette-Pitre, N. (2009). Building a virtual learning community of problem solvers: example of CASMI community. *ZDM, 41*(1–2), 245–256.

Freiman, V., & Manuel, D. (2015). Relating students' perceptions of interest and difficulty to the richness of mathematical problems posted on the CAMI website. *Quadrant, 25*(2), 61–84.

Freiman, V. (2010). Mathematically gifted students in inclusive settings: Case of New Brunswick, Canada. In B. Sriraman & K. H. Lee (Eds.), *Elements of creativity and giftedness in mathematics*. (pp. 161–172). Sense Publishers.

Freiman, V., & Volkov, A. (2004). Early mathematical giftedness and its social context: The cases of Imperial China and Soviet Russia. *Journal of the Korean Society of Mathematical Education Series D: Research in Mathematical Education, 8*, 157–173.

Gagné, F. Y. (2003). *Giftedness in early childhood* (3rd ed.).

Gagné, F. (2009). Building gifts into talents: Detailed overview of the DMGT 2.0. In B. MacFarlane & T. Stambaugh (Eds.), *Leading change in gifted education: The Festschrift of Dr. Joyce Vantassel-Baska* (pp. 61–80). Waco, TX: Prufrock Press.

Gavin, M. K., Casa, T. M., Adelson, J. L., & Firmender, J. M. (2013). The impact of advanced geometry and measurement units on the achievement of grade 2 students. *Journal of Research in Mathematics Education, 44*(3), 478–510.

Gavin, M. K., Casa, T. M., Adelson, J. L., Carroll, S. R., & Sheffield, L. J. (2009). The impact of advanced curriculum on the achievement of mathematically promising elementary students. *Gifted Child Quarterly, 53*, 188–202.

Geake, J. G. (2006). Mathematical brains. *Gifted and Talented, 10*(1), 2–7.

Geake, J. G. (2009). *The brain at school: Educational neuroscience in the classroom*. Sydney: McGraw Hill & Open Univ. Press.

Glaser, R. (1988). Cognitive science and education. *International Social Science Journal, 115*, 21–45.

Gorodetsky, M., & Klavirb, R. (2003). What can we learn from how gifted/average pupils describe their processes of problem solving? *Learning and Instruction, 13*(3), 305–325.

Greenes, C., & Mode, M. (1999). Empowering teachers to discover, challenge and support students with mathematical promise. In L. Sheffield (Ed.), *Developing mathematically promising students* (pp. 121–132). Reston, VA: NCTM.

Greenes, C., Teuscher, D., & Regis, T. P. (2010). Preparing teachers for mathematically talented middle school students. In M. Saul, S. Assouline, & L. J. Sheffield (Eds.), *The peak in the middle: Developing mathematically gifted students in the middle grades* (pp. 77–91). Reston, VA: NCTM.

Gross, M. U. M. (2003). *Exceptionally gifted children* (2nd ed.). London: Routledge.

Harrison, C. (2003). Giftedness in early childhood: The search for complexity and connection. *Roeper Review, 26*(2), 78–84.

Heller, K., & Ziegler, A. (Eds.). (2007). *Begabt sein in Deutschland*. Berlin: LIT Verlag.

Hoeflinger, M. (1998). Developing mathematically promising students. *Roeper Review, 20*(4), 244–247.

Holton, D., Cheung, K., Kesianye, S., Falk de Losada, M., Leikin, R., Makrides, G., et al. (2009). Teacher development and mathematical challenge. In E. Barbeau & P. Taylor (Eds.), *Challenging mathematics in and beyond the classroom (The 16th ICMI Study)* (pp. 205–242). New York: Springer.

Hong, E., & Aqui, Y. (2004). Cognitive and motivational characteristics of adolescents gifted in mathematics: Comparisons among students with different types of giftedness. *Gifted Child Quarterly, 48*, 191–201.

Hoyles, C. (2001). Steering between skills and creativity: A role for the computer? *For the Learning of Mathematics, 21*, 33–39.

Irvine, S. H., & Berry, J. W. (1988). The abilities of mankind: A revaluation. In S. H. Irvine & J. W. Berry (Eds.), *Human abilities in cultural context* (pp. 3–59). Cambridge Univ. Press.

Jensen, E. (2000). *Brain-based learning*. San Diego, CA: The Brain Store.

Jensen (Sheffield), L. R. (1973). *The relationships among mathematical creativity, numerical aptitude, and mathematical achievement.* Unpubl. dissertation. Austin, TX: The Univ. of Texas at Austin.

Jin, S. H., Kim, S. Y., Park, K. H., & Lee, K. J. (2007). Differences in EEG between gifted and average students: Neural complexity and functional cluster analysis. *International Journal of Neuroscience, 117*, 1167–1184.

Johnsen, S., & Sheffield, L. J. (Eds.). (2012). *Using the common core state standards for mathematics with gifted and advanced learners.* Washington, DC: NAGC.

Käpnick, F. (1998). Mathematisch begabte Kinder. Modelle, empirische Studien und Förderungsprojekte für das Grundschulalter. Frankfurt am Main.

Karp, A. (2016). A brief history of specialized mathematics schools. In B. Vogeli (Ed.), *Special secondary schools for the mathematically talented: An international panorama.* Hackensack, NJ: World Scientific.

Karsenty, R., & Friedlander, A. (2008). Professional development of teachers of mathematically gifted students: An agenda under construction. In R. Leikin (Ed.), *MCG5 Proceedings* (pp. 454–456). Haifa, Israel: Univ. of Haifa.

Kattou, M., Christou, C., & Pitta-Pantazi, D. (2015). Mathematical creativity or general creativity? In K. Krainer & N. Vondrová (Eds.), *Proceedings of CERME9.* Prague, Czech Republic: Charles University and ERME.

Krutetskii, V. A. (1976). *The psychology of mathematical abilities in schoolchildren.* Chicago: Univ. of Chicago Press.

Leblanc, M., & Freiman, V. (2011). Mathematical and didactical enrichment for pre-service teachers: Mentoring online problem solving in the CASMI project. *The Montana Mathematics Enthusiast, 8*(1–2), 291–318.

Lee, K. H., Choi, Y. Y., Gray, J. R., Cho, S. H., Chae, J. H., & Lee, S. (2006). Neural correlates of superior intelligence: Stronger recruitment of posterior parietal cortex. *NeuroImage, 29*, 578–586.

Leikin, M., Paz-Baruch, N., & Leikin, R. (2013). Memory abilities in generally gifted and excelling-in-mathematics adolescents. *Intelligence, 41*, 566–578.

Leikin, R., & Kloss, Y. (2011). Mathematical creativity of 8th and 10th grade students. In *CERME7 Proceedings* (pp. 1084–1093). Univ. of Rzeszów, Poland: ERME.

Leikin, R., & Winicky-Landman, G. (2001). Defining as a vehicle for professional development of secondary school mathematics teachers. *The Mathematics Education Research Journal, 3*, 62–73.

Leikin, R. (2009). Exploring mathematical creativity using multiple solution tasks. In R. Leikin, A. Berman, & B. Koichu (Eds.), *Creativity in mathematics and the education of gifted students* (pp. 129–145). Rotterdam: Sense Publishers.

Leikin, R. (2011). The education of mathematically gifted students: Some complexities and questions. *The Mathematics Enthusiast, 8*(1–9). Retrieved from http://scholarworks.umt.edu/tme/vol8/iss1/9

Leikin, R. (2013). Evaluating mathematical creativity: the interplay between multiplicity and insight. *Psychological Test and Assessment Modeling, 55*(4), 385–400.

Mhlolo, M. K. (2014). Opening up conversations on the plight of the mathematically talented students in sub-saharan African countries. In *Proceedings of the 8th Int. MCG Conference.* Denver, CO. Retrieved January 24, 2016, from http://igmcg.edu.haifa.ac.il/Conferences

Milgram, R., & Hong, E. (2009). Talent loss in mathematics: Causes and solutions. In R. Leikin, A. Berman, & B. Koichu (Eds.), *Creativity in mathematics and the education of gifted students* (pp. 149–161). Rotterdam: Sense Publishers.

Mönks, F. J. & Pflüger, R. (2005). *Gifted education in 21 European countries: Inventory and perspective. Radboud University Nijmegen.* https://www.bmbf.de/pub/gifted_education_21_eu_countries.pdf

NAGC and CEC. (2013). *NAGC—CEC Teacher Preparation Standards in Gifted and Talented Education.* Retrieved January 24, 2016, from http://www.nagc.org/sites/default/files/standards/NAGC-%20CEC%20CAEP%20standards%20%282013%20final%29.pdf

NAGC and CEC. (2014). *Knowledge and skill standards in gifted education for all teachers.* Retrieved January 24, 2016, from http://www.nagc.org/resources-publications/resources/national-standards-gifted-and-talented-education/knowledge-and

National Council of Teachers of Mathematics (NCTM). (1980). *An agenda for action: Recommendations for school mathematics of the 1980s.* Reston, VA: NCTM.

National Council of Teachers of Mathematics (NCTM). (1995). Report of the NCTM task force on the mathematically promising. *NCTM News Bulletin, 32.*

National Governors Association (NGA) Center for Best Practices, Council of Chief State School Officers (CCSSO). (2010). *Common core state standards for mathematics.* Washington D.C.: National Governors Association Center for Best Practices, Council of Chief State School Officers.

National Research Council. (2001). *Adding it up: Helping children learn mathematics.* In J. Kilpatrick, J. Swafford, & B. Findell (Eds.), Mathematics learning study committee, center for education, division of behavioral and social sciences and education. Washington, DC: Nat. Acad. Press.

Nolte, M. (2012). Mathematically gifted young children—Questions about the development of mathematical giftedness. In H. Stöger, A. Aljughaiman, & B. Harder (Eds.), *Talent development and excellence* (pp. 155–176). Berlin, London: Lit Verlag.

Nolte, M. (2013). Twice exceptional children: Mathematically gifted children in primary schools with special needs. In *CERME 8 Proceedings.* Ankara: Middle East Technical Univ.

Nordheimer, S., & Brandl, M. (2016). Students with hearing impairment: Challenges facing the identification of mathematical giftedness. In K. Krainer & N. Vondrová (Eds.), *CERME9 Proceedings* (pp. 1032–1038). Prague, Czech Republic: Charles University and ERME.

National Science Board (NSB). (2010). Preparing the next generation of STEM innovators: Identifying and developing our nation's human capital. (NSB-10-33). Washington, DC: NSF.

O'Boyle, M. W. (2008). Mathematically gifted children: Developmental brain characteristics and their prognosis for well-being. *Roeper Review, 30,* 181–186.

O'Boyle, M. W. (2005). Some current findings on brain characteristics of the mathematically gifted adolescent. *International Education Journal, 6*(2), 247–251.

O'Boyle, M. W., Cunnington, R., Silk, T., Vaughan, D., Jackson, G., Syngeniotis, A., & Egan, G. (2005). Mathematically gifted male adolescents activate a unique brain network during mental rotation. *Cognitive Brain Research, 25,* 583–587.

Olszewski-Kubilius, P. (2013, May 20). Setting the record straight on ability grouping. *Education Week.* Retrieved January 24, 2016, from http://www.edweek.org/tm/articles/2013/05/20

Öystein, H. P. (2011). What characterizes high achieving students' mathematical reasoning? In B. Sriraman & K. H. Lee (Eds.), *The elements of creativity and giftedness in mathematics* (pp. 193–216). Rotterdam: Sense.

Pelczer, I., Singer, F. M., & Voica, C. (2013). Cognitive framing: A case in problem posing. *Procedia— SBS,* PSIWORLD 2012, *78,* 195–199.

Pelczer, I., Singer, F. M., & Voica, C. (2014). Improving problem-posing capacities through inservice teacher training programs: challenges and limits. In *PME38 Proceedings* (pp. 401–408). Vancouver, Canada.

Presenti, M., Zargo, L., Crivello, F., Mellet, E., Samson, D., & Duroux, B. (2001). Mental calculation in a prodigy is sustained by right prefrontal and medial temporal areas. *Nature Neuroscience, 4,* 103–107.

Printer, C. P., Moon, T. R., & Brighton, C. M. (2015). Characteristics of students' mathematical promise when engaging with problem-based learning units in primary classrooms. *Journal of Advanced Academics, 26*(1), 24–58.

Raghubar, K. P., Barnes, M. A., & Hecht, S. A. (2010). Working memory and mathematics: A review of developmental, individual difference, and cognitive approaches. *Learning and Individual Differences, 20,* 110–122.

Reis, S. M., Westberg, K. L., Kulikowich, J. M., & Purcell, J. H. (1998). Curriculum compacting and achievement test scores: What does the research say? *Gifted Child Quarterly, 42,* 123–129.

Renzulli, J. S. (1986). The three-ring conception of giftedness: A developmental model for creative productivity. In R. J. Sternberg & J. E. Davidson (Eds.), *Conceptions of giftedness* (pp. 53–92). Cambridge, UK: Cambridge UP.

Renzulli, J. S. (2012). Reexamining the role of gifted education and talent development for the 21st century: A four-part theoretical approach. *Gifted Child Quarterly, 56*(3), 150–159.

Sheffield, L. J. (2003). *Extending the challenge in mathematics: Developing mathematical promise in K—8 pupils*. Thousand Oaks, CA: Corwin Press.

Sheffield, L. J. (2006). Developing mathematical promise and creativity. *Journal of the Korea Society of Mathematical Education Series D: Research in Mathematical Education, 10*(1), 1–11.

Sheffield, L. J. (2009). Developing mathematical creativity: Questions may be the answer. In R. Leikin, A. Berman, & B. Koichu (Eds.), *Creativity in mathematics and the education of gifted students*. Rotterdam, The Netherlands: Sense Publishers.

Sheffield, L. J., Bennett, J., Berriozabal, M., DeArmond, M., & Wertheimer, R. (1999). Report of the NCTM task force on the mathematically promising. In L. J. Sheffield (Ed.), *Developing mathematically promising students* (pp. 309–316). Reston, VA: NCTM.

Sheffield, L. J., Firmender, J., Gavin, M. K., & Casa, T. M. (2012). Project M^2: Mentoring young mathematicians. *MCG7 Proceedings* (pp. 269–276). MCG: Busan, Korea.

Singer, F. M. (2007). Beyond conceptual change: Using representations to integrate domain-specific structural models in learning mathematics. *Mind, Brain, and Education, 1*(2), 84–97.

Singer, F. M. (2009). The Dynamic infrastructure of mind—A hypothesis and some of its applications. *New Ideas in Psychology, 27*(1), 48–74.

Singer, F. M., Ellerton, N., & Cai, J. (2013). Problem-posing research in mathematics education: New questions and directions. *Educational Studies in Mathematics, 83*(1), 1–7.

Singer, F. M., Ellerton, N. F., & Cai, J. (Eds.). (2015). *Mathematical problem posing: From research to effective practice*. New York: Springer.

Singer, M., & Sarivan, L. (2009). Curriculum reframed: MI and new routes to teaching and learning in Romanian universities. In J. Q. Chen, S. Moran, & H. Gardner (Eds.), *Multiple intelligences around the world* (pp. 230–244). New York: Wiley.

Singer, F. M., & Sarivan, L. (2011). Masterprof: A program to educate Teachers for the Knowledge Society. In F.M. Singer & L. Sarivan (Eds.), *Procedia—Social and Behavioral Sciences*, 11 (2011), p. 7–11.

Singer, F. M., & Singer, B. (2010). STAR: A collaborative technology for personalized feedback in learning. In P. Escudeiro (Ed.), *Proceedings of 9th European Conference on e-Learning*, (pp. 554–560). Reading, UK: Academic Publishing Limited.

Singer, F. M., & Voica, C. (2015). Is problem posing a tool for identifying and developing mathematical creativity? In F. M. Singer, N. Ellerton, & J. Cai (Eds.), *Mathematical problem posing: From research to effective practice* (pp. 141–174). New York: Springer.

Singer, F. M., & Voica, C. (2016). When mathematics meets real objects: how does creativity interact with expertise in problem solving and posing? In R. Leikin & B. Sriraman (Eds.), *Creativity and giftedness: interdisciplinary perspectives*. New York: Springer.

Sriraman, B. (2005). Are giftedness and creativity synonyms in mathematics? An analysis of constructs within the professional and school realms. *Journal of Secondary Gifted Education, 17*(1), 20–36.

Stepanak, J. (1999). Meeting the needs of gifted students: Differentiating mathematics and science instruction. USA: Northwest Regional Educational Laboratory *The differentiation toolbox* (2009) KUDs. Retrieved from http://people.virginia.edu/~mws6u/diff/index.htm

Subotnik, R. F., Robinson, A., Callahan, C. M., & Gubbins, E. J. (2012). *Malleable minds: Translating insights from psychology and neuroscience to gifted education*. Storrs: Univ. of Connecticut, NRCGT.

Szabo, A. (2015). Mathematical problem-solving by high achieving students: Interaction of mathematical abilities and the role of the mathematical memory. In K. Krainer & N. Vondrová (Eds.), *Proceedings of CERME9* (pp. 1087–1093). Prague, Czech Republic: Charles University and ERME.

Tall, D. (2008). The transition to formal thinking in mathematics. *Mathematics Education Research Journal, 20*(2), 5–24.

Tomlinson, C. A. (1999). *The differentiated classroom: Responding to the needs of all learners.* Alexandria, VA: ASCD.

Usiskin, Z. (2000). The development into the mathematically talented. *The Journal of Secondary Gifted Education, 11*, 152–162.

Vogeli, B. R., & Karp, A. (Eds.). (2003). *Activating mathematical talent.* Denver, CO: NCSM.

Vogeli, B. R. (Ed.). (2016). *Special secondary schools for the mathematically talented: An international panorama.* Singapore: World Scientific.

Vilkomir, T., & O' Donoghue, J. (2009). Using components of mathematical ability for initial development and identification of mathematically promising students. *International Journal of Mathematical Education in Science and Technology, 40*(2), 183–199.

Voica, C., & Singer, F. M. (2013). Problem modification as a tool for detecting cognitive flexibility in school children. *ZDM, 45*(2), 267–279.

Voica, C., & Singer, F. M. (2014). Problem posing: A pathway to identifying gifted students. In *MCG8 Proceedings* (pp. 119–124). Univ. of Denver, Colorado, USA.

Winkler, S., & Brandl, M. (2016). Process-based analysis of mathematically gifted pupils in a regular class at primary school. In K. Krainer & N. Vondrová (Eds.), *CERME9 Proceedings* (pp. 1101–1102). Prague, Czech Republic: Charles University, Faculty of Ed. and ERME.

Winner, E. (2000). The origins and ends of giftedness. *American Psych., 55*(1), 159–169.

Ziegler, A. (2005). The actiotope model of giftedness. In R. J. Steinberg & J. E. Davidson (Eds.), *Conceptions of giftedness* (2nd ed., pp. 422–443). Cambridge, U.K.: Cambridge Univ. Press.

Zmood, S. (2014). Fostering the promise of high achieving mathematics students through curriculum differentiation. In J. Anderson, M. Cavanagh, & A. Prescott (Eds.), *Curriculum in focus: Research guided practice (Proceedings of the 37th annual conference of the Mathematics Education Research Group of Australasia)* (pp. 677–684). Sydney: MERGA.

www.ingramcontent.com/pod-product-compliance
Ingram Content Group UK Ltd.
Pitfield, Milton Keynes, MK11 3LW, UK
UKHW020216231225
466357UK00011B/177